ARROYO CENTER

Army Installation Rail Operations

Implications of Increased Outsourcing

Ellen M. Pint, Beth E. Lachman, Jeremy M. Eckhause, Steven Deane-Shinbrot

For more information on this publication, visit www.rand.org/t/RR2009

Preface

This document reports the results of a research project entitled Army Rail and Public-Private Partnerships. As part of this project, we compared the Army's three business models for installation rail operations: government owned, government operated (GOGO); government owned, contractor operated (GOCO); and privatized. We also developed a methodology to compare the costs and risks of the business models to help the Army determine whether it could rely to a greater extent on privatized rail operations.

This document describes the results of our analysis, which is based on detailed information gathered through visits to one installation of each type and data on Army rail requirements, costs, and performance at installations with deployable units. We estimated potential savings from privatization and determined possible risk factors, such as a decrease in responsiveness to short-notice deployments, loss of surge capacity, likelihood of accidents or violations of safety and environmental rules, and unexpected cost increases.

This research was sponsored by the Deputy Chief of Staff, G-4, and conducted within the RAND Arroyo Center's Forces and Logistics Program. RAND Arroyo Center, part of the RAND Corporation, is a federally funded research and development center sponsored by the U.S. Army.

The Project Unique Identification Code (PUIC) for the project that produced this document is HQD157609.

Contents

Figures

Tables

Summary

Rail is one of the most efficient means for transporting heavy equipment over long distances, but Army demand for rail is highly variable, consisting primarily of shipments to ports for deployments and to combat training centers for exercises. Thus, Army-owned rail assets and rail crews are not always fully utilized. The Army shipped a total of approximately 20,000 loaded rail cars at a cost of $120 million in fiscal year 2015, compared to nearly 30 million rail cars carried and $70 billion in revenue for U.S. Class I railroads. The Army relies on commercial rail carriers for off-post rail movements, but it currently has three business models for on-post rail operations:[1] government owned, government operated (GOGO); government owned, contractor operated (GOCO); and privatized.

The Office of the Deputy Chief of Staff, G-4, asked RAND Arroyo Center to evaluate the three business models and determine whether greater reliance on commercial rail assets could meet Army rail needs at a lower cost. As part of this research, we gathered data on Army rail requirements, costs, and performance at installations with deployable units and visited one installation of each type to obtain insights on the differences between the business models. We then developed an approach to compare the costs and risks of the three business models across installations.

Overview of Business Models

The Army's business models for installation rail operations vary primarily in two ways: the ownership of the locomotives that are used for on-site positioning of rail cars and the personnel who perform these duties. Under the GOGO business model, the Army purchases and maintains its own locomotives and employs government civilians to operate the locomotives and conduct other rail-related activities. Under the GOCO model, the Army also purchases and maintains locomotives, but the rail crews are contractor employees. In most cases, they are hired under a larger umbrella contract for Logistics Readiness Center (LRC) operations on the installation. Under the privatized model, both the locomotives and the rail crews are provided by the rail carrier as part of the shipping costs for the off-post movement.

However, some aspects of installation rail operations do not vary by business model. First, in all three cases, the Army owns the track on its installations, and the track and other rail infrastructure are maintained by the installation's Directorate of Public Works. Second, the

---

[1]   Army installation rail operations include positioning rail cars for loading and unloading equipment, training and assisting unit personnel who load equipment onto rail cars, and assembling trains and positioning them for pickup by commercial rail carriers.

rail cars utilized are a combination of government-owned and commercial cars, depending on the size and weight of the equipment and containers being shipped. Third, all off-post movements are handled by commercial rail carriers.

Based on our interviews with installation rail personnel, the GOGO and GOCO business models are seen as more flexible and responsive to the schedules of Army units, whereas the rail carriers must fit Army requirements into their commercial work schedules. In addition, GOGO and GOCO rail crews and locomotives are immediately available to begin preparations for a movement in the event of a short-notice deployment. However, the GOGO and GOCO models have higher fixed costs for locomotives and rail crews. Under the privatized model, the Army only pays these costs as part of the costs of a shipment. Both GOGO and GOCO rail crews can perform other transportation-related duties when they are not busy with rail operations, and they have low turnover rates, so they are familiar with the installation rail infrastructure and standard operating procedures. The primary distinction between the GOGO and GOCO models is that contractor employees are easier than government employees to hire and fire in response to changes in workload. In addition, when an unexpected deployment occurs, the Army can demand overtime and other duties for government civilian employees, which could be more difficult under the GOCO model, depending on the contract terms and conditions.

Cost and Risk Analysis

The first step in our analysis is to determine the circumstances under which the privatized model is likely to be less expensive than the GOGO and GOCO business models. At installations where there are potential cost savings, we then consider the potential risks of privatization and steps that could be taken to mitigate those risks.

Essentially, our cost analysis involves a comparison of the higher fixed costs of the GOGO and GOCO business models with the higher variable costs of the privatized model. Since the commercial rail carriers do not itemize their bills, we are reliant on an estimate that the costs of the additional services they provide under the privatized model range between $400 and $900 per rail car. When compared with the fixed costs of paying full-time rail crews and maintaining locomotives, we find that installations that ship an average of fewer than 30 to 40 rail cars per month are likely to have lower costs under the privatized model. However, the annual cost savings are modest, most likely less than $300,000 per year at each installation.

The potential cost savings from privatization must be weighed against possible risks. First, the Army will be dependent on commercial rail carriers to provide locomotives and rail crews in a timely manner to load equipment and containers. While it should be possible to obtain these resources for planned movements, such as training exercises, they may not be readily available for short-notice deployments unless contingency plans are in place, because the rail carriers' capacity is optimized for their commercial business, and the Army is an infrequent customer. Second, the Army may lose some internal surge capacity, because GOGO and GOCO installations share rail crews when they need to conduct 24-hour operations to load a brigade combat team's equipment for a deployment or training rotation. As a result, the Army would need to obtain additional surge capacity from commercial carriers. Third, there may be an increased risk of accidents or violations of safety and environmental rules if commercial rail crews are unfamiliar with the installation's rail infrastructure. Fourth, there may be unexpected additional costs. The Army currently contracts separately for each rail movement, and some commercial rail carriers have local monopoly power to raise their prices after privatization. In addition, some infrastructure improvements may be needed to accommodate commercial locomotives or meet Federal Railroad Administration requirements on Army installations.

Some of these risks can be mitigated by specifying contract terms and conditions with rail carriers before privatizing installation rail operations. Long-term contracts with rail carriers should specify the cost and availability of rail crews and locomotives for peacetime and surge operations, as well as any requirements for rail crews, such as familiarity with installation rail infrastructure, background checks, and adherence to safety, security, and environmental rules. In addition, contracts should specify the costs that will be charged for additional services, including switching rail cars and overseeing the loading and tying down of equipment by Army unit personnel.

Recommendations

At installations with low rail activity rates where privatization may be cost-effective, the Army must balance potential savings against the risks of privatization. Factors to be considered include the installation's deployment requirements, whether it shares rail crews with other installations, whether it is served by a local monopoly rail carrier, and whether infrastructure investments that could offset potential cost savings will be required.

The recompetition of Fort Sill's GOCO LRC contract in 2017 offers the Army an opportunity to conduct a more detailed comparison of the costs and risks of privatization. It should require the bidders for the LRC contract to separately specify the costs of GOCO rail operations, which could then be compared with a second solicitation to commercial rail carriers for bids to provide privatized operations. These bids should include the cost and availability of rail crews and locomotives; prenegotiated rates for rail movements, including additional services; and any required infrastructure investments.

Acknowledgments

We thank our action officers, COL Lawrence Kominiak and COL William Shinn, for their guidance throughout this research project. We are also grateful for the assistance of Julie Nato and James Cloe in obtaining data and scheduling installation visits. In addition, we thank the LRC personnel and rail crews at Forts Hood, Sill, Benning, and Irwin and the Marine Corps Logistics Base Barstow for hosting our visits and generously providing their time to answer our questions. We also thank staff members at Fort Drum, Military Surface Deployment and Distribution Command, the U.S. Army Audit Agency, the Association of American Railroads, Burlington Northern Santa Fe, CSX, and Norfolk Southern who participated in telephone interviews and provided data.

We thank our RAND colleagues Marc Robbins, Patricia Boren, Elvira Loredo, and Bruce Held for their guidance and assistance on this project. Craig Bond of RAND and Ryan Sullivan of the Naval Postgraduate School provided thoughtful reviews that have improved the quality of this document. In addition, we appreciate the administrative assistance of Rosie Velasquez and Joan Myers.

Abbreviations

AAA	U.S. Army Audit Agency
ABCT	armored brigade combat team
ADA	air defense artillery
AFSB	Army Field Support Brigade
BCT	brigade combat team
BNSF	Burlington Northern Santa Fe
CAB	combat aviation brigade
CN	Canadian National
CP	Canadian Pacific
CTC	combat training center
DGRC	Defense Non-Tactical Generator and Rail Equipment Repair Center
DoD	Department of Defense
DODX	Department of Defense rail cars
DPW	Directorate of Public Works
ETA	estimated time of arrival
FRA	Federal Railroad Administration
FY	fiscal year
GIS	geographic information system
GOCO	government owned, contractor operated
GOGO	government owned, government operated
HEMTT	Heavy Expanded Mobility Tactical Truck
IBCT	infantry brigade combat team
IBCT-A	airborne infantry brigade combat team
ITRS	installation transportation rail services
JRTC	Joint Readiness Training Center
KCS	Kansas City Southern

LEAD	Letterkenny Army Depot
LIDA	Letterkenny Industrial Development Authority
LRC	Logistics Readiness Center
MCLB	Marine Corps Logistics Base
MOA	memorandum of agreement
NS	Norfolk Southern
NTC	National Training Center
OMB	Office of Management and Budget
PLS	Palletized Load System
PPP	power projection platform
PWS	Performance Work Statement
QTS	Quality Transportation Services
RC	reserve component
SB	sustainment brigade
SBCT	Stryker Brigade Combat Team
SDDC	U.S. Military Surface Deployment and Distribution Command
SLWC	Stillwater Central
TEA	Transportation Engineering Agency
TTX	commercial rail cars
UP	Union Pacific
WG	wage grade
WSOR	Wisconsin & Southern Railroad

CHAPTER ONE

Introduction

Rail is one of the most efficient means for transporting heavy equipment, yet Army rail operations face a number of challenges. Demand for rail is highly variable, including contingency deployments, rotational deployments, combat training center (CTC) rotations, and equipment distribution movements, so Army-owned rail assets are not always fully utilized. In fiscal year (FY) 2015, the Army shipped a total of about 20,000 loaded rail cars at a cost of $120 million, compared to nearly 30 million rail cars carried and $70 billion in revenue for U.S. Class I railroads (Association of American Railroads, 2016).[1] Moreover, existing Army, Department of Defense (DoD), and private-sector rail assets used to move Army equipment, such as locomotives and rail cars, are aging and many need to be modernized. The Army may be able to make more cost-effective use of rail transportation by building on existing relationships with the major U.S. freight railroads and increasing its reliance on private-sector rail assets that it only uses when needed. However, the Army must also ensure that these rail assets are available for short-notice surge and contingency operations.

The Army currently has three different business models for installation rail operations:[2] government owned, government operated (GOGO); government owned, contractor operated (GOCO); and privatized. Under the GOGO business model, the Army owns locomotives, track, loading ramps, container-handling equipment, and other assets located on the installation. Rail crews, typically consisting of a locomotive engineer, conductor, and brakeman, are government employees. Their duties include switching rail cars on the installations and overseeing loading and unloading of equipment by soldiers. Under the GOCO business model, the Army owns locomotives and other equipment and infrastructure, but a contractor provides rail crews that conduct switching operations and oversee loading and unloading of equipment. The rail crews may be provided under a standalone contract with a local rail carrier, or as part of an umbrella contract to operate the installation's Logistics Readiness Center (LRC). The terms and conditions of these contracts are not standardized across installations.

Under the privatized business model, the Army still owns rail infrastructure, but it does not station any locomotives on the installation or have permanent rail crews. Instead, when the installation transportation office plans a rail movement, it requests that the rail carrier provide additional services, including lining up the rail cars in the correct order for loading,

---

[1] Installations with deployable units that are included in this study accounted for about 14,000 of the rail cars shipped and $86 million of off-post shipment costs in FY 2015.

[2] Army installation rail operations include positioning rail cars for loading and unloading equipment, training and assisting unit personnel who load equipment onto rail cars, and assembling trains and positioning them for pickup by commercial rail carriers. All off-post rail movements are performed by commercial rail carriers.

and overseeing the loading and tying down of equipment by soldiers. Under all three business models, commercial rail carriers conduct all off-installation movements after the trains have been loaded, using their own locomotives and a mixture of government-owned (DODX) and commercial (TTX) rail cars.

The Office of the Deputy Chief of Staff, G-4, asked RAND Arroyo Center to evaluate the three business models and determine whether greater reliance on commercial rail assets could meet Army rail needs at a lower cost. The research tasks included

- reviewing and evaluating GOGO rail facilities and operations at Fort Hood, Texas
- reviewing and evaluating GOCO rail facilities and operations at Fort Sill, Oklahoma
- reviewing and evaluating privatized rail facilities and operations at Fort Drum, New York, and/or Fort Benning, Georgia
- comparing and contrasting alternative business models and identifying findings or methodology that can be generalized to other Army installations.

Research Methodology

As part of this research, we gathered data on Army rail requirements, costs, and performance at the target installations, including installation rail services contracts and agreements, rail deployment studies conducted by the Transportation Engineering Agency (TEA), rail shipment data collected by the Military Surface Deployment and Distribution Command (SDDC), and records of rail movements maintained by LRC personnel at the target installations. We also leveraged data collected by the U.S. Army Audit Agency (AAA) as part of an audit of Army installation rail operations conducted in 2015.[3] In addition to the target installations, we collected data on other U.S. Army installations with deployable units, as shown in Table 1.1.

Second, we visited one installation of each type—Fort Hood, Fort Sill, and Fort Benning—and conducted a telephone interview with LRC personnel at Fort Drum. At each of these installations, we interviewed government and contractor personnel (if applicable) to obtain their perspectives on how the business model functioned on their installations, and the advantages and disadvantages of each business model. We also visited Fort Irwin, California, and the Marine Corps Logistics Base (MCLB) at Yermo, California, to learn more about rail operations supporting the National Training Center (NTC), because it has more rail activity than any other U.S. installation.[4] To get the perspectives of commercial rail carriers, we also conducted telephone interviews with personnel from the Association of American Railroads, Norfolk Southern (NS), CSX, and Burlington Northern Santa Fe (BNSF).

Third, we developed an approach to compare the costs and risks of the three business models across installations. This approach can be applied to all the installations listed in Table 1.1, based on available data.

---

[3] See U.S. Army Audit Agency, 2017. We also conducted a telephone interview with Army auditors.

[4] Rail operations at MCLB Barstow's Yermo Annex Railyard are described in Appendix A.

Table 1.1
Army Installations by Rail Business Model

GOGO	GOCO	Privatized
Fort Hood	Fort Sill	Fort Benning
Fort Carson	Fort Bliss	Fort Drum
Joint Base Lewis-McChord	Fort Bragg	Fort Knox
Fort Leonard Wood	Fort Campbell	Fort Polk
Fort Riley	Fort Irwin	
Fort Stewart	Fort McCoy	

SOURCE: U.S. Army Audit Agency, 2017.

Outline of This Report

The remainder of this report is organized as follows. Chapter Two provides a more detailed comparison of the three business models and describes installation rail operations at the three target installations. In Chapter Three, we present our cost and risk analysis methodology, apply it using data on Fort Sill and Fort Hood, and discuss how it can be generalized to other installations. Chapter Four summarizes our results and recommendations.

Business Models for Army Installation Rail Operations

In this chapter, we provide a more detailed comparison of the three business models employed at Army installations that were evaluated in this study: GOGO, GOCO, and privatized installation rail operations. We also illustrate how the business models operate at our target installations—i.e., GOGO operations at Fort Hood, GOCO operations at Fort Sill, and privatized operations at Fort Benning and Fort Drum.

Overview and Comparison of Business Models

In the context of Army rail assets and operations, the business model varies primarily in two ways: the ownership of the locomotives that are used for on-site positioning and maneuvering of rail cars and the personnel utilized to perform such duties. Under the GOGO business model, the Army purchases and maintains its own locomotives and employs Department of the Army civilians to operate the locomotives and conduct other on-post rail activities. The GOCO model is largely identical to the GOGO model, but the rail crews are contractor personnel rather than government employees. Often these contract employees are hired under a larger umbrella contract for LRC operations on the installation. For the privatized model, both the locomotives and the rail crews responsible for on-post movements are provided by the rail carrier (often under a subcontract with a local rail services provider) from whom that installation has purchased additional services as part of the shipping costs for the off-post movement.

It is important to understand that there are several aspects of Army rail operations that do not vary by business model. First, under all three business models, the Army owns the track on its installations, and the installation Directorate of Public Works (DPW) and its personnel (which can be government or contractor employees) are responsible for maintaining the rail infrastructure; this applies even under a privatized model, in which the Army neither owns any locomotives nor has any full-time rail crews at the installation.[1] Second, the rail cars utilized for off-post movements are a combination of DODX and TTX rail cars, depending on availability, requirements, and costs to position them from other locations. Third, all off-post movements of trains for any purpose (training, deployments, transferring heavy equipment, etc.) are handled by commercial rail carriers, regardless of the business model used at the installation. Depending on the origin, destination, and volume, these rail shipments may be handled by multiple commercial rail carriers.

---

[1]   Note that this aspect of privatized rail operations differs from other types of privatization, in which the Army sells or enters into long-term leases for infrastructure, such as housing and utilities.

Figure 2.1 summarizes the areas of overlap and differences among the three business models for Army rail.

The personnel who perform various tasks in preparing for an Army rail movement can also differ by business model. Figure 2.2 summarizes the roles of different personnel, based on the three installations we visited. Army unit personnel and unit movement officers (who were all government civilians at the three installations, even when other LRC operations were contracted out) conduct predeployment preparation activities, such as determining the number and types of equipment that will be transported and preparing them for loading. Unit personnel are also responsible for loading their equipment and securing it onto the rail cars, but these operations are overseen by a combination of government civilians, LRC contractor personnel, and rail carrier personnel, depending on the business model. The personnel who line up the rail

Figure 2.1
Key Rail Assets and Operations That Vary by Business Model

	Privatized	**GOCO**	**GOGO**
Rail crews, inspectors, and within-post rail car movements	Provided by rail carrier as part of shipping costs	Contractor personnel	Army civilian personnel
Locomotives		Army purchased and maintained	Army purchased and maintained
Installation-owned track	Maintained by DPW personnel and/or contractors		
Rail cars	Combination of DODX and commercial cars		
Off-post movements	Provided by commercial rail carriers		

RAND *RR2009A-2.1*

Figure 2.2
Who Performs the Tasks for Each Phase of Installation Rail Deployment

Phases of Rail Deployment	Fort Benning (Privatized)	Fort Sill (GOCO)	Fort Hood (GOGO)
Predeployment preparation activities	Army units and government civilians	Army units and government civilians	Army units and government civilians
Loading and tying down of equipment on rail cars	Army units, government civilians, and contractors	Army units, government civilians, and contractors	Army units and government civilians
Rail car switching and train assembly on the installation	NS	Contractor employees	Government civilians
Line-haul train movement over commercial rail lines	NS	SLWC, BNSF, or UP	BNSF

NOTE: BNSF = Burlington Northern Santa Fe; NS = Norfolk Southern, SLWC = Stillwater Central, and UP = Union Pacific

RAND *RR2009A-2.2*

cars at the loading ramps in the appropriate order for the loading of equipment and containers and then assemble the cars into trains for off-post movement also vary by business model. At Fort Benning, these tasks are performed by NS (or a subcontractor) as additional services under the movement contract; at Fort Sill, they are performed by LRC contractor personnel; and at Fort Hood, by government civilian personnel.

Line-haul movements are performed by the rail carriers who own or have the rights to use the commercial rail lines leading onto the installation. When an off-post rail movement is needed, the installation transportation office sends a DD Form 1085 (Domestic Freight Routing Request and Order) to SDDC, which then advertises the movement to eligible rail carriers, who submit bids. Fort Benning and Fort Hood have monopoly rail carriers, NS and BNSF, respectively. Fort Sill is served directly by a short line, Stillwater Central (SLWC), which connects with two major carriers, BNSF and Union Pacific (UP). If Army rail movements cross regions served by different carriers, the cars are interchanged and the prices charged by each carrier are incorporated into the originating carrier's bid.

Table 2.1 provides an overview of the rail activities and other features of the installations we examined in detail for this study, based on data gathered by the AAA audit, SDDC rail shipment data, and data provided by the installations we visited. Most data are from FY 2015. Installation rail personnel we interviewed said that most GOGO versus GOCO decisions were made in the 1980s and 1990s as a result of public-private competitions for LRC operations (then known as Directorates of Logistics) conducted under the rules specified in the Office of Management and Budget (OMB) Circular A-76.[2] LRC operations tended to be outsourced at installations that primarily had a training mission or lighter units (infantry or airborne),

Table 2.1
Rail Activity at Target Installations

Rail Feature	Fort Benning	Fort Drum	Fort Sill	Fort Hood
Business model	Privatized	Privatized	GOCO	GOGO
Number of locomotives	0	0	4	8[a]
Average locomotive operating hours per month	N/A	N/A	16	229
Maximum locomotive operating hours per month	N/A	N/A	22	694
Total rail cars moved on post (loaded and empty)	333	717	1,859	79,592
Monthly average rail cars shipped off post (loaded only)	25	34	15	168
Local carrier	NS	CSX	SLWC, BNSF, UP	BNSF

SOURCES: U.S. Army Audit Agency, 2017, SDDC rail shipment data.

[a]AAA study indicated eight locomotives, but only six were stationed at Fort Hood when we visited in May 2016.

---

2   OMB Circular A-76 (OMB, 2003) describes the rules and procedures that must be followed for public-private competitions to convert any work performed by government civilian personnel. The best contractor bid is compared with an in-house proposal called a Most Efficient Organization, and the workload is converted to contract if it would result in cost savings of at least 10 percent. However, a moratorium on DoD public-private competitions has been in effect since the passage of section 325 of the National Defense Authorization Act for FY 2010. The moratorium has been extended by subsequent legislation. See, for example, Assistant Secretary of Defense for Manpower and Reserve Affairs, 2016.

whereas those with armored units tended to remain GOGO.[3] However, there are exceptions to this pattern, such as Fort Bliss, which had a training mission in the 1990s but now houses armored units. Privatization tended to occur at installations with relatively low rail activity rates, where the fixed costs of rail crews and locomotives were more likely to outweigh the higher variable costs of contracting with rail carriers for additional services.

In the remainder of this chapter, we will describe rail operations at each of the installations shown in Table 2.1.

Fort Hood Rail Operations: Government-Owned, Government-Operated Model

As noted earlier, Fort Hood is an example of the GOGO business model, so it has Army civilian rail crews operating several Army-owned locomotives. It is one of the busiest Army installations in terms of rail activity, providing support for the training and deployment of four brigade combat teams (BCTs), as well as other units, including engineer, chemical, air defense artillery, military police, and military intelligence brigades. When Fort Hood deploys units for combat operations, rail is used to transport tanks and other heavy equipment to a nearby port, usually the Port of Beaumont, Texas[4] (see Figure 2.3). Fort Hood has two three-person rail crews (engineer, conductor, and brakeman), along with five blocking and bracing inspectors.[5] In addition to conducting rail operations and training Army unit personnel on proper procedures for loading and unloading equipment on rail cars, rail crews are cross-trained for other transportation functions, such as material handling and container management, when not performing rail duties.

As of May 2016, the installation had six locomotives that crews used in pairs to position and move rail cars loaded with heavy equipment and to deal with track grade as high as 2 percent. Four of these locomotives are new Genset engines (see Figure 2.4), and the other two are older GP40 engines. According to Fort Hood rail personnel, Genset engines are highly computerized systems that were fielded without sufficient training. They require specialized diagnostic and repair equipment and break down more frequently than the GP40 engines. Personnel also said that most of the time, at least one engine is down awaiting repair, and sometimes they have to wait months for service personnel from the Defense Non-Tactical Generator and Rail Equipment Repair Center (DGRC), located at Hill Air Force Base.[6] Fort Hood has 28 miles of active track and two rail yards, which are maintained by the installation DPW (as with all installations, regardless of business model).

---

[3] One partial reason for this outcome is that U.S. Army Training and Doctrine command conducted a nationwide A-76 review of all in-house Directorates of Public Works and Logistics beginning in 1997. It created a centralized study-management approach that helped ensure that all A-76 studies were completed within the required four-year timeline, thus avoiding cancellations. See National Council for Public-Private Partnerships, 2002.

[4] In 2014, the Port of Beaumont was the fourth-busiest shipping port in the United States. See Bureau of Transportation Statistics, undated.

[5] One rail crew position and three blocking and bracing inspectors are funded by Overseas Contingency Operations budgets.

[6] For more information on the DGRC, see Bacchus, 2012. We also heard that other installations, such as Fort Bliss, had difficulties operating and maintaining Genset engines. However, Fort Sill rail personnel told us that they did not have many repair issues with their Genset engines because they are second-generation Genset engines, whereas Fort Hood has first-generation Genset engines.

Figure 2.3
Map of Rail Lines Between Fort Hood and Port of Beaumont, Texas

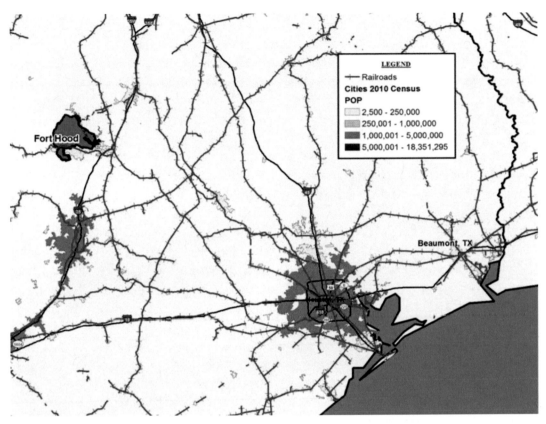

SOURCE: RAND-generated GIS map.
RAND RR2009A-2.3

Figure 2.4
Two Genset Locomotives at Fort Hood, Texas

SOURCE: Photo by Jeremy Eckhause.
RAND RR2009A-2.4

Based on data maintained by Fort Hood rail personnel, rail crews performed nearly 80,000 rail car movements in FY 2015. The vast majority (roughly 90 percent) were on-post movements, primarily positioning rail cars for loading, unloading, storage, and training. Since Army unit personnel must be trained on loading, tying down, and unloading their equipment from rail cars, a few rail cars are moved for such purposes. In addition to the large number of internal movements, the number of off-post movements is significant. Fort Hood personnel recorded on-post movements of rail cars on 167 days in FY 2015 (including more than half of weekdays), and there were almost as many days with off-post movements (121 days). These off-post movements can include sending or returning equipment from a training exercise or deployment, the receipt or shipment of empty DODX or TTX cars, or the shipment or receipt of individual pieces of equipment too heavy or large to be sent via line-haul truck.

A scheduled movement of an armored brigade combat team (ABCT) or other large shipment from Fort Hood requires coordination with Quality Transportation Services (QTS), an SDDC contractor who coordinates requests for rail cars from the DODX and TTX fleets, as well as the local commercial rail carrier, BNSF. For a deployment or training exercise, the request is typically provided to QTS 30 to 45 days in advance to allow for the positioning of rail cars. BNSF needs about one week's notice for normal movements, or 4 to 5 days for surge operations. The trains are loaded by Army unit personnel with oversight by the blocking and bracing inspectors. Once assembled, the trains are positioned near the edge of the post by the Fort Hood crews for off-post transportation by BNSF. Fort Hood rail personnel said that they have a good relationship with BNSF, but they expressed some concerns about its responsiveness. BNSF has a local rail monopoly, and military shipments account for less than 1 percent of its business. Since Fort Hood is an infrequent customer in comparison with commercial customers, which tend to have more frequent, regularly scheduled shipments, BNSF must fit Army shipments into its normal commercial schedule, and the exact time of pickup can vary by several hours.

Despite the presence of GOGO rail crews, Fort Hood does not have enough personnel to maintain 24/7 rail operations for a sustained period due to safety regulations.[7] However, installations within the 407th Army Field Support Brigade (AFSB) region share crews to assist each other with major unit movements. It takes approximately 10 days to load the six to eight trains needed to move an ABCT to the NTC or port, and possibly 6 to 8 days longer if Fort Hood does not receive assistance from other installations, such as Fort Sill or Fort Campbell. During recent deployments, Fort Hood was also supported by soldiers from a reserve component (RC) rail unit, but its funding for training has been reduced, so the RC rail crews will not be able to maintain their certifications. Without certification, the RC rail personnel can advise but not assist with rail loading and unloading operations.

Fort Hood has the ability to store approximately 700 to 800 rail cars on post, using both the modernized west rail yard and the older east rail yard. This capability could be reduced to 500 rail cars should the east rail yard be removed to allow for additional housing and other installation facilities.[8] Figure 2.5 shows a map of the track and rail yards (shown in red) on Fort Hood. The west rail yard, where the majority of rail activity occurs, is on the left side of

---

[7]   Rail crews are limited to 12-hour shifts for no more than six consecutive days. Thus, two crews can only conduct 24-hour operations for six days.

[8]   According to Fort Hood LRC personnel, the decision to remove the east rail yard was made in the 1990s, after the west rail yard was built. However, the track has not yet been removed, so it was used to store rail cars during the early deployments for Operation Iraqi Freedom and has also been used for rail car maintenance activities.

Figure 2.5
Map of Fort Hood Track (in red)

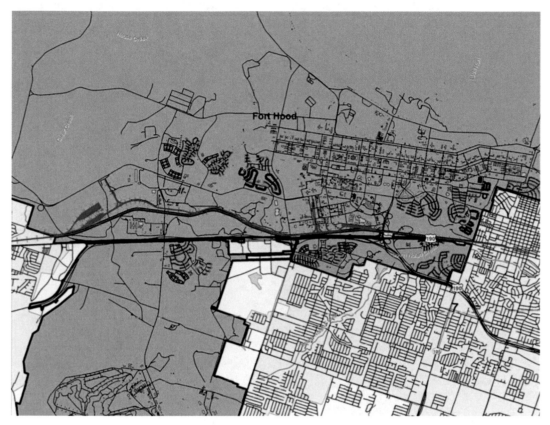

SOURCE: RAND-generated GIS map.
RAND *RR2009A-2.5*

the map. The upper set of parallel tracks is used to load rail cars; the lower set is used for rail car storage. The east rail yard is located in the center right of the map.

Based on our discussions with Fort Hood LRC personnel, there are several advantages to the GOGO model, especially in cases such as Fort Hood, where the volume of rail traffic is quite high. The model generally lends itself to very little turnover in the crews, reducing transition costs and likely increasing safety on post because crews are familiar with the installation's rail infrastructure. Most of the GOGO crews have additional expertise outside their core position, allowing for flexibility and responsiveness when other positions are short-staffed. When an unexpected deployment occurs, the Army can demand overtime and other duties for government civilians, which could be more difficult under the GOCO or privatized models due to contract terms and conditions[9] or competing business incentives. On the other hand, the ability to hire quickly or temporarily under a GOGO model is limited,[10] and the fixed costs of

---

[9] For example, Fort Hood personnel said that installations were not allowed to put surge requirements into LRC contracts unless they could be documented in a detailed specification that contractors could bid on.

[10] Fort Hood personnel noted that it was sometimes difficult to hire qualified rail crew members. The training process for new hires who are not fully qualified could take up to two years, and employees sometimes left for higher-paying, private-sector jobs after obtaining qualifications.

an underutilized crew (if not appropriately or feasibly cross-trained) can be quite high, though the high volume of rail activity at Fort Hood tends to mitigate this risk. Additionally, the fully burdened ownership costs of locomotives can be substantial if they are not heavily utilized, though leases could mitigate the risk of midterm uncertainty in the demand for rail operations.

Fort Sill Rail Operations: Government-Owned, Contractor-Operated Model

Fort Sill railroad operations are an example of the GOCO business model. As under the GOGO business model, the installation has Army-owned locomotives, but the rail crews are contractor employees. At the time of our visit to Fort Sill in July 2016, the installation had two cross-functional rail crews. Each crew consisted of one engineer, one conductor, and one brakeman, and they were cross-trained to operate the locomotives, to perform basic locomotive maintenance functions, and to train and oversee Army unit personnel on rail car loading. Fort Sill had recently hired a second rail crew to help support other installations in the 407th AFSB region, such as Fort Hood and Fort Carson, when they need additional crews for major rail movements to the NTC or a port. Fort Sill LRC personnel stated that it was easier for Fort Sill to hire additional qualified contractor employees than for GOGO installations to hire government employees.

The GOCO rail crews are provided under an umbrella contract with Primus for Fort Sill's LRC operations. As specified in the contract, Primus is responsible for railway operations and equipment management and maintenance, including switching and spotting rail cars; maintaining daily contact with the commercial rail carriers; moving rail cars between the loading and unloading area and the interchange with the main rail line; maintaining a record of every rail car switched in and out of Fort Sill; and performing daily operator maintenance and quarterly organizational maintenance and repair of railway equipment. Rail operations occupy only about 50 percent of the Fort Sill rail crews' time. When they are not operating or maintaining locomotives, the crews perform other freight functions, including loading trucks and rail cars, helping with unit movements, and operating material-handling equipment.

The freight chief is a contractor employee who is responsible for truck and rail carrier oversight. However, some LRC senior managers are government civilian employees, such as the chief of unit movements. He provides oversight during truck and rail car loading and unloading. Under the terms of the contract, he cannot directly supervise the rail contractor employees, but he can provide advice and suggestions. The contract expires in 2017, and the recompetition may offer an opportunity for the Army to compare the costs of the Fort Sill GOCO business model with privatization.

Fort Sill's rail assets include four locomotives, three relatively new Gensets, and one older 120-ton locomotive. Unlike Fort Hood, Fort Sill rail crews said they have not had many major maintenance issues with their Genset engines because they are second-generation Gensets, whereas Fort Hood has first-generation Gensets. Fort Sill rail personnel also told us that they did not need the older engine given their level of rail activity. Based on SDDC data, Fort Sill shipped 178 loaded rail cars in FY 2015, or an average of about 15 per month.

Fort Sill has 16 miles of track and one rail yard. The installation also has a fairly new turning loop that was built in 2012 at a cost of about $3.5 million. It was constructed to reduce the time needed to reorient rail cars. Previously, rail crews had to use a wye located six miles off post, which took about four hours each time. As noted previously, Fort Sill is served directly by

Figure 2.6
Map of the Rail Lines Between Fort Sill and Surrounding Cities

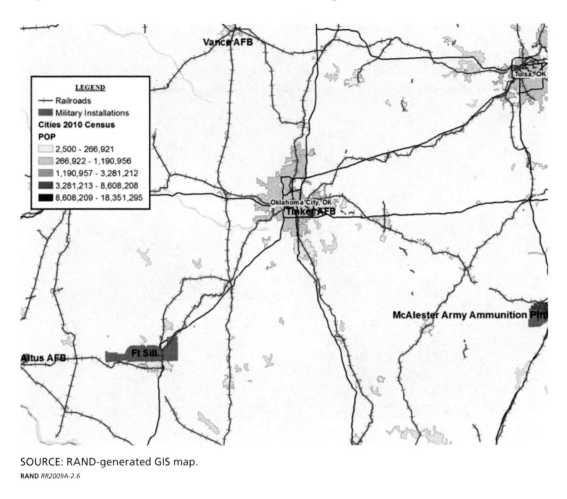

SOURCE: RAND-generated GIS map.
RAND *RR2009A-2.6*

SLWC (a short-line rail carrier), which connects with two major rail carriers, UP and BNSF. Usually, SLWC provides service to Oklahoma City, Altus, or Tulsa, where the rail cars are picked up by BNSF or UP (see Figure 2.6).[11]

In June 2016, Fort Sill's rail operations were affected by a natural disaster, providing an example of one of the risks to Army installation rail operations that is not related to the installation's rail business model, since the track is still owned and maintained by the Army. Heavy rains and flooding washed out the ground under the track leading to a culvert on the only rail corridor into and out of Fort Sill's rail yard (see Figure 2.7 for a photograph of the damage). The damaged track is located between Fort Sill's rail yard and the installation boundary, so at the time of our visit, Fort Sill rail crews could not use their locomotives to position rail cars for loading or move any trains to meet the rail carrier. Until the damage is repaired, Fort Sill will need to use portable loading ramps located near the edge of the installation and rely on the commercial rail carrier to position the rail cars. Rail personnel estimated that this workaround is likely to cause the process of loading and unloading trains to take twice as long, and it thus

---

[11] SDDC TEA personnel noted that BNSF locomotives and crews do not pick up trains at Fort Sill, and UP rarely does so. If rail operations at Fort Sill are privatized, the most likely provider would be SLWC.

Figure 2.7
Photo of Fort Sill Track Damage from 2016 Flood

SOURCE: Photo by Beth Lachman.
RAND RR2009A-2.7

could affect deployment time lines if a short-notice contingency occurs. Figure 2.8 shows the approximate location of the track damage and the portable loading ramps.

According to Fort Sill DPW personnel, the estimated cost for repairs is $2.5 million and the estimated time to complete the repair is one year. The repairs are estimated to take so long because of the time required to acquire the funding and schedule the repair.[12] Fort Sill DPW personnel thought it would take about 30 days to repair the track once the money has been acquired and a contract is in place. This example illustrates the need for coordination across Army organizations to maintain rail capabilities, due to the separation of responsibilities and funding between the LRC, which manages rail operations, and the DPW, which manages track inspection, maintenance, and repair. However, privatization of rail operations would not affect the DPW's responsibility for maintenance of track, loading ramps, and other facilities, and the Army would still incur the costs of maintaining and repairing the track if it privatized rail operations at Fort Sill.

Fort Sill government civilian personnel said that they were satisfied with the GOCO business model. They thought it was more flexible than the GOGO model, because it is easier for

---

[12] Fort Sill personnel said that the Army has an in-house track repair team at McAlester Army Ammunition Plant that could design and perform the repairs, but it is usually booked up long in advance. The alternative is to contract for the repairs, but in that case, the DPW would need to provide a design.

Figure 2.8
Map Showing the Approximate Location of the Fort Sill Track Damage and the Portable Loading Area

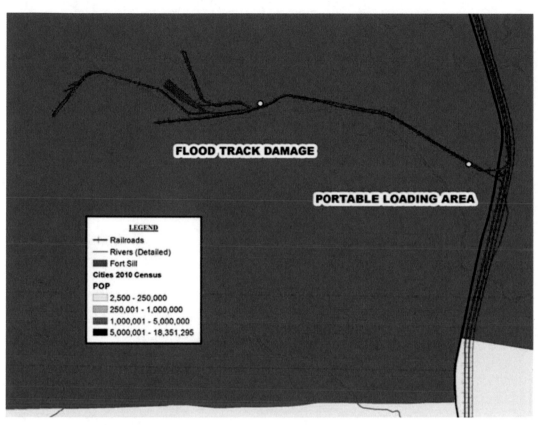

SOURCE: RAND-generated GIS map.
RAND *RR2009A-2.8*

the contractor to hire and fire personnel based on the workload. However, under the GOGO model, the Army has more control over the workforce. The GOCO model is also more flexible than privatization, because the GOCO crew can adapt to the unit's time line, whereas the commercial rail carriers must balance the Army's demands with those of their other customers and thus may not be as responsive. The GOCO crews are also flexible in the sense that they can assist with other freight operations when not needed for rail. Although the LRC contract is periodically recompeted, the rail crews typically have the right of first refusal to keep their positions if a new contractor wins the competition. Thus, Fort Sill had not experienced any problems with high turnover among its rail crews.[13]

Fort Benning and Fort Drum Rail Operations: Privatized Model

In this section, we focus primarily on Fort Benning, but we also provide some information on privatized rail operations at Fort Drum.

---

[13] In fact, one contractor employee was preparing to retire after 28 years of service and had been employed at Fort Sill since its LRC operations were first outsourced in 1988.

Fort Benning

Fort Benning's railroad operations are privatized, which means that the installation has no locomotives and relies on its commercial rail carrier for any rail car movements and oversight of Army unit personnel when they load and tie down equipment. NS is Fort Benning's local monopoly rail carrier. Fort Benning's freight movement staff and LRC contractor employees reported that they have a good working relationship with NS, which is important since they have no locomotives of their own to perform internal rail car movements. Figure 2.9 shows an overview of the rail lines on and around Fort Benning.

Most of Fort Benning's LRC operations are performed by VS2[14] under an Enhanced Army Global Logistics Enterprise contract. The LRC's Transportation Division has nine government positions, but only two of them have rail responsibilities: the division head and a logistic management specialist in the Unit Movement Coordination office. The division head has a small oversight role with respect to rail. The logistic management specialist manages unit movements for Fort Benning, including truck, air, and rail movements. The Transportation Division also has eight contractor employees that support freight movements, mostly by

Figure 2.9
Map of the Rail Lines on and near Fort Benning

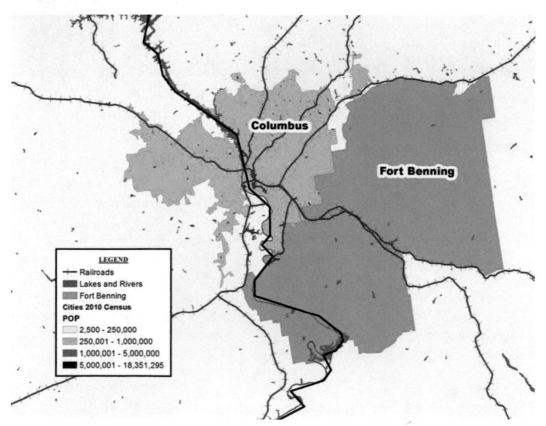

SOURCE: RAND-generated GIS map.
NOTE: The GIS data source shows a rail line heading south into the Fort Benning main cantonment area that is no longer in service.
RAND RR2009A-2.9

---

[14] VS2 is a joint venture of VSE Corporation and CB&I Federal Services. The contract was awarded in May 2015 after two previous awards were protested. See Department of Defense Inspector General, 2014; and Tomkins, 2015.

truck but also including some rail and air movements. They manage all the military freight[15] for the post, from large ground movements to small packages.

Fort Benning shipped a total of 298 loaded rail cars in FY 2015, averaging about 25 rail cars per month. Rail personnel reported that, in the past, they had a busy year if they had four rail movements, with trains going out twice a year and coming back twice a year. The contractor employee who manages the rail yard said that she spent less than 10 percent of her time on rail in 2015. However, during a rail movement, she spends about 80 percent of her time on the movement. She also trains Army unit personnel on rail loading procedures. Rail training used to be an annual requirement, but in recent years, they have had some problems ensuring that units receive the training. She also works closely with NS to ensure that the rail cars are lined up in the correct order for loading, since they do not have their own engines to switch rail cars.

Fort Benning has a total of 6.35 miles of track and one rail yard. Over the last several years, Fort Benning expanded its rail infrastructure to improve its deployment capability. In particular, the railhead lacked a bypass, so it took the commercial rail carrier three to four hours to switch trains in and out of the railhead. In 2009–2010, Fort Benning started a rail expansion project that cost $7.4 million. The project was completed in April 2014, adding a bypass, five extra storage tracks, and an extra 1,980-foot ramp and spur (a total of 3.87 miles of track). When we visited Fort Benning in August 2016, it had the ability to load 183 rail cars with 420 pieces of equipment and containers per 24-hour period and store up to 350 rail cars (including the bypass).

In 2015, Fort Benning's ABCT was downsized to a maneuver battalion task force as part of the Army's reduction in end strength, so it will have less demand for peacetime rail services in the future.[16] However, installation rail personnel said that if the Georgia National Guard's 48th Infantry BCT (IBCT) were to deploy, they would likely use Fort Benning's rail facilities. Fort Benning also has placed over 100 of the ABCT's buildings in caretaker status in anticipation that the installation might grow at some time in the future. Thus, it could potentially house a BCT if the Army were to grow again, which would increase its need for rail.

Fort Drum

Fort Drum is also an example of the privatized business model. Its local rail carrier is CSX. Fort Drum shipped a total of 407 rail cars in FY 2015, or an average of about 34 per month. It has two IBCTs, which typically require four 65-car trains to deploy to a port or send to a Joint Readiness Training Center (JRTC) rotation. They usually request a shipment at least two weeks in advance, primarily to ensure that the appropriate rail cars are available. Most of an IBCT's equipment can be shipped on commercial rail cars, but some heavier equipment, such as the Heavy Expanded Mobility Tactical Truck (HEMTT) and Palletized Load System (PLS), require DODX cars.

Since Fort Drum does not have a locomotive, LRC personnel must request the specific rail car types needed, and CSX delivers them in solid groups of DODX or TTX rail cars. They have to be loaded on separate ramps, because they are different heights. CSX charges for additional services, such as providing a locomotive and crew for switching and technical support to make sure that the soldiers load and tie down the equipment properly. However, these services are not priced separately, so it is not clear exactly how much of the cost is related to additional

---

[15] Military freight does not include any movement of personal goods for soldiers.

[16] Fort Benning's primary mission is training. It houses the Maneuver Center of Excellence and provides basic training and One Station Unit Training for infantry and armor.

services versus other factors, such as distance and weight of the shipment. Fort Drum person-nel said that, occasionally, it would be useful to have a locomotive to reposition cars to unload a deadlined vehicle or to reduce the amount of time they need to pay for a CSX crew, but since they only do a few movements a year, it would probably not be cost-effective. If they had a locomotive, they would also need an engine house and other infrastructure.[17]

There is a 40-car minimum to get a dedicated train. If a unit needs to ship at least that much equipment, it will go by rail. If the unit has less than that, they prefer to send the equipment by truck. It will take much longer for the equipment to go "manifest" (i.e., as part of a larger train assembled with rail cars from other customers) instead of by dedicated train. For example, it takes 5 to 7 days for a dedicated train to get to Fort Polk but 15 to 17 days if it goes manifest.

Like other privatized installations, Fort Drum still owns its track, which is maintained by two WG-8 government civilian personnel employed by the DPW. Fort Drum personnel said that their rail was refurbished in 2009–2012 and the track weight was increased to 115 lbs. However, other facilities have not been upgraded. For example, there is a two-lane road leading up to the loading area, which gets very congested. There is also inadequate lighting on the road for nighttime loading, although there is good lighting over the track.

Strengths and Weaknesses of the Privatized Model

The privatized model has the advantage of little to no fixed overhead costs for locomotives or crews, so for locations with infrequent rail activity, this approach can provide significant savings even though the rail carriers charge additional fees per rail car to provide a rail crew and locomotive for switching and other services. As Fort Benning and Fort Drum typically have only a handful of movements per year (roughly three to four), the privatized model is likely cost-effective, particularly since, at both installations, the local rail carrier is receptive to requests for on-post movements. For example, NS was described as usually very responsive and typically less of a bottleneck than the internal DoD procurement processes for requesting a shipment. In addition to savings from not employing full-time rail crews, these installations can also avoid the ownership costs of locomotives, which creates significant savings if on-post movements are rare. Finally, with a privatized model, it could potentially be easier to surge with additional commercial crews since the business relationship with the local carrier already exists, whereas obtaining crews from the local rail carrier might not be possible at an instal-lation with a GOGO or GOCO model that does not habitually require additional services.

The success of the privatized model depends not only on infrequent rail activity to jus-tify the higher marginal prices for each shipment but also on an effective relationship with the local carrier. A less responsive carrier may not be able to meet the installation's time lines for deployment or training requirements. Due to the specific location of each installation within its regional rail network, competition from other carriers, and other demands for their services, local rail carriers may more or less actively seek business from the Army. Understanding these dynamics for each installation is critical to making a decision to use a privatized model, as the risk for not meeting the requirements could be significant.[18]

---

[17] A less expensive alternative might be a trackmobile or rail car spotter. SDDC TEA personnel indicated that the Army is conducting proof-of-principal evaluations to determine the utility of this alternative for limited applications.

[18] For example, personnel from SDDC TEA noted that NS and CSX have rail yards supporting local freight service located near Fort Benning and Fort Drum, respectively. Thus, on-post switching is compatible with those rail carriers' local busi-ness models. In contrast, Fort Hood and Fort Stewart are located much farther away from their carriers' rail yards, so on-post switching would not be a good fit with the carriers' local business models.

CHAPTER THREE

Cost and Risk Analysis

In this chapter, we describe the cost and risk analysis that we used to consider whether the Army should change rail business models at installations with deployable units. It is important to recognize that such decisions need to be based on comparing the potential cost savings with the risks to Army installation rail operations and readiness that could arise from changing the business model at a given installation. First, we discuss the cost analysis, then the risks to rail operations, and finally ways to address the tradeoff between the risk and cost concerns.

To set the context for the cost and risk analysis, we first describe the criteria for successful installation rail operations. The most important criterion is the timely delivery of equipment in peacetime and surge operations to ensure that the Army maintains readiness and meets deployment time lines. Second, the Army needs the capability to surge rail loading operations for deployments. This objective includes the availability of potentially scarce resources such as the crews, locomotives, and rail cars needed to perform the required movements. In addition, having diversity, redundancy, and robustness in the Army's rail infrastructure helps ensure the Army can meet its operational and deployment missions given uncertainties about future deployment needs. A third important factor for successful installation rail operations is the familiarity of the rail crews with the installation rail infrastructure. Such familiarity is needed to improve the efficiency of rail operations and reduce the risks of accidents and derailments. Locomotives are highly specialized vehicles that must be operated differently based on rail facility characteristics, such as the slope of the installation terrain and the location of switches.[1] The fourth criterion is being able to provide installation rail services in a cost-effective manner for both steady state and surge operations. Ideally, this would involve reducing peacetime costs while maintaining the capability to surge to meet short-notice deployment requirements.

Estimated Cost Differences Between Army Installation Rail Operation Business Models

In this section we present an analysis of the costs associated with changing installation rail operations business models. We start by examining a case study for assessing the estimated costs of changing the Fort Sill GOCO model to a privatized model. Based on information obtained from LRC personnel, Fort Sill currently has two GOCO rail crews, or a total of

---

[1] For example, Fort Hood has a section of track with a 2 percent grade. Rail personnel there told us that less experienced locomotive engineers and other rail crew members who come to Fort Hood must be educated on additional safety precautions to effectively operate a locomotive and secure rail cars under such conditions.

six personnel, costing approximately $200,000 per year, assuming that they spend half their time on rail operations. We assume that they spend the remainder of their time assisting with other freight operations but that Fort Sill could hire lower-cost employees (who do not have rail training and qualifications) if rail operations were privatized. Fort Sill has four locomotives with annual maintenance costs of $146,000, based on information obtained from the AAA audit. Thus, the total annual fixed costs of GOCO rail operations at Fort Sill are approximately $346,000.

We weigh that against the additional costs charged by commercial rail carriers to provide services such as switching rail cars and overseeing loading operations to ensure that soldiers are properly loading and tying down equipment on the rail cars. Personnel at MCLB Barstow estimated that, based on their experience, these services cost between $400 and $900 per rail car.[2] Therefore, we estimate that the break-even point, where the costs of the GOCO and privatized business models would be equal, is 866 rail cars shipped per year (or 72 per month) if the cost is $400 per car, and 385 rail cars shipped per year (32 per month) if the cost is $900 per car. Figure 3.1 compares the total annual costs of GOCO rail operations with those of privatized rail operations at the two different price points. The slopes of the lines are based on Fort Sill's average rail shipment costs in FY 2014–2015 of $4,823 per car.[3]

Since Fort Sill shipped about 15 cars per month during FY 2015, our cost calculations suggest that there may be some potential cost savings from privatization of rail operations at Fort Sill. However, as Figure 3.1 indicates, these cost differences are not very large. At Fort Sill's FY 2015 rail activity rate, estimated annual savings from privatization would range between about $185,000, if the commercial rail carriers charged $900 per car for additional services, and about $275,000, if they charged $400 per car. We must also consider the potential risks from privatization, which we discuss in greater detail in the next section.

Our second example compares the annual costs of GOGO and privatized rail operations at Fort Hood. Based on information obtained from LRC personnel and the AAA audit, Fort Hood currently has two government civilian rail crews costing approximately $418,000 per year, assuming that engineers and conductors are wage grade 9 (WG9) and brakemen are WG7. We also assume that Fort Hood rail crews spend 100 percent of their time on rail operations, based on the higher rail activity rates at Fort Hood. As of May 2016, Fort Hood had six locomotives, with annual maintenance costs of $219,000. Thus, the total annual fixed costs of GOGO rail operations at Fort Hood are approximately $637,000. We estimate that the break-even point between GOGO and privatized rail operations at Fort Hood ranges from 1,592 cars per year (133 per month), if rail carriers charge $400 per rail car for additional services, and 708 cars per year (59 per month), if rail carriers charge $900 per car. Since Fort Hood shipped an average of 168 cars per month in FY 2015, the current GOGO model is likely less expensive than privatization.

Figure 3.2 compares the total annual costs of GOGO rail operations with those of privatized rail operations at the two different price points. The slopes of the lines are based on Fort Hood's

---

[2]   We attempted to use SDDC data on rail shipment costs, weights, and distances to estimate the differences in prices charged to installations by business model but did not find any statistically significant differences, due to errors in the data and other confounding factors. However, we did find that installations served by more than one commercial rail carrier paid significantly lower prices than those with a local monopoly rail carrier.

[3]   For shipments during this period with complete data, Fort Sill paid $1,253,900 to ship 260 rail cars. The primary destinations were Fort Irwin; the Port of Beaumont, Texas; Fort Bliss; and Anniston Army Depot. Note that this value identifies the slopes for the lines in Figure 3.1 but does not affect the crossing points.

Figure 3.1
Cost Comparison of Business Models at Fort Sill

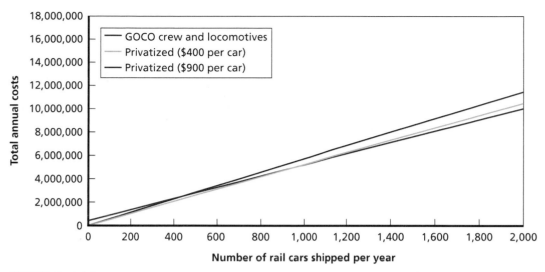

SOURCE: Cost data.
RAND RR2009A-3.1

Figure 3.2
Cost Comparison of Business Models at Fort Hood

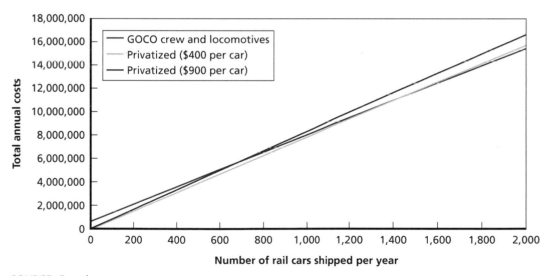

SOURCE: Cost data.
RAND RR2009A-3.2

average rail shipment costs in FY 2014–2015 of $7,406 per car.[4] Again, these cost differences are not very large. At Fort Hood's FY 2015 rail activity rate, estimated annual savings from retaining the GOGO business model would range between about $170,000, if the commercial rail carriers charged $400 per car for additional services, and about $1,178,000, if they charged $900 per car.

---

[4]   For shipments during this period with complete data, Fort Hood paid $19,180,541 to ship 2,590 rail cars. The primary destinations were Fort Irwin; the Port of Beaumont, Texas; Joint Base Lewis McChord; Fort Riley; and Sierra Army Depot. Note that Fort Hood has a local monopoly rail carrier, whereas Fort Sill is served by competing rail carriers.

Table 3.1 shows the effects of varying some of our assumptions on the break-even point between the GOCO and privatized business models, based on local factors such as wage rates, number of rail crews, fraction of time spent on rail operations, and number of locomotives. We also show the break-even points at a cost of $650 per car for additional services (i.e., the midpoint between $400 and $900 per car). In the upper part of the table, we vary one assumption at a time, holding the other assumptions the same as those for the Fort Sill base case. In the last two rows of the table, we first combine all the assumptions more favorable to privatization, then we combine all the assumptions less favorable to privatization, in order to obtain upper and lower bounds on the break-even points. A higher break-even point favors privatization, because the installation must have a higher rail activity rate to make GOCO rail operations less costly than privatization.

For example, the national median wages for locomotive engineers, conductors, and brakemen are considerably higher than those reported by Fort Sill, so an installation reimbursing a contractor at these wage rates would have higher fixed costs for the GOCO business model and would need to ship a larger number of rail cars per year to be cost-effective relative to privatization. Similarly, if we include a share of the capital costs of replacing locomotives (e.g., $1.8 million for a Genset with a 30-year life-span), it would also raise the fixed costs of the GOCO business model and the break-even point with privatization. The recommendations of the AAA audit, such as reducing the number of locomotives at each installation, would tend to lower the fixed costs of the GOGO and GOCO business models and make them more attractive relative to privatization.

Based on the cost analysis, it would appear that installations shipping fewer than 30 to 40 rail cars per month are most likely to have cost savings from privatization. Figure 3.3 shows the average monthly rail cars shipped in FY 2015 by each of the installations in our sample, excluding the CTCs, which tend to have much higher rail activity rates. Using a threshold of 30 cars per month, the best candidates for privatization appear to be Fort Bragg,

Table 3.1
Effects of Assumptions on Break-Even Point Between Government-Owned, Contractor-Operated Model and Privatization

GOCO Model Assumptions	Break-Even Point with Privatization		
	$900 per Car	$650 per Car	$400 per Car
Base case (Fort Sill)	32	44	72
National median wages for rail crew	44	61	99
Crew spends 100% of time on rail	42	58	95
Crew spends 25% of time on rail	27	37	61
One three-person rail crew	23	32	52
Including locomotive replacement costs	54	75	122
Reduce to three locomotives	29	40	65
Combined Assumptions			
Most favorable to privatization	76	106	172
Least favorable to privatization	20	27	44

Figure 3.3
Fiscal Year 2015 Average Monthly Number of Loaded Rail Cars Shipped by Installation

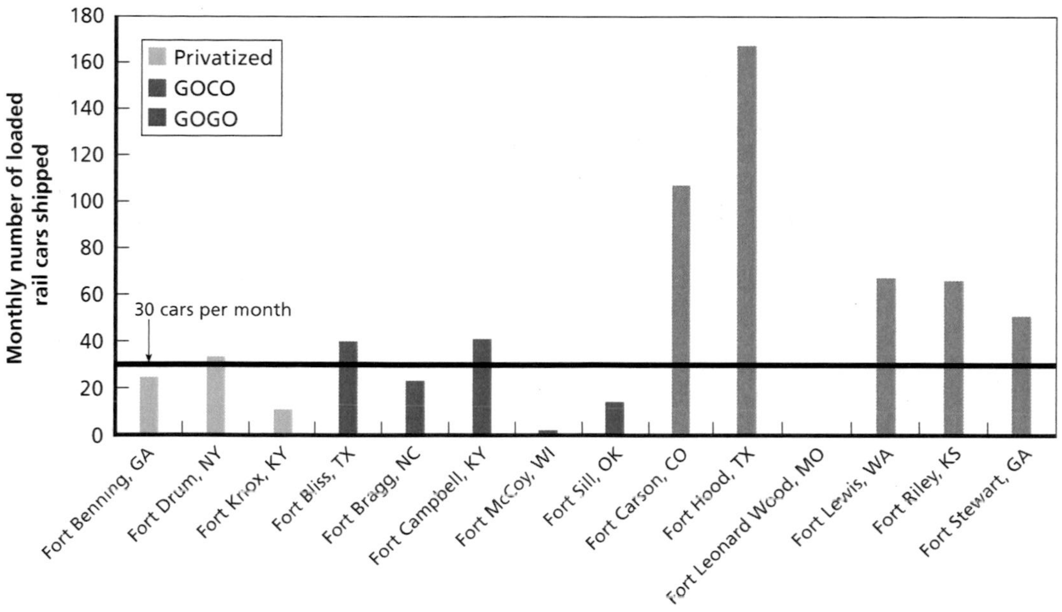

SOURCE: SDDC rail shipment data.
NOTE: Fort Leonard Wood shipped no loaded rail cars in FY15. A less complete SDDC data source indicates that it shipped at least 37 loaded rail cars (an average of 3 per month) in FY14.
RAND RR2009A 3.3

Fort Leonard Wood, Fort McCoy, and Fort Sill. However, the Army must also consider the potential risks from privatization, which we discuss in greater detail in the following section.

Risks to Army Installation Rail Operations from Changing Business Models

Before making a decision to change the rail operations business model at an installation to reduce peacetime costs, the Army must also consider possible risks—i.e., whether the new business model will be able to meet the criteria for successful installation rail operations. In other words, the potential cost savings must be weighed against the potential risks from privatization. First, we discuss the main types of risks that could arise from reducing the availability of Army rail crews and locomotives, then we discuss some additional risks posed by privatization that were mentioned during our interviews.

We group the main risks of reducing Army rail assets into four categories, analogous to the criteria for successful rail operations:

- delayed rail movements that could affect Army readiness or deployment requirements
- loss of surge capacity for deployments
- increased likelihood of accidents or violations of safety and environmental rules
- unexpected additional costs.

These risks are interrelated, but for discussion purposes we explain each type of risk separately with some illustrative examples.

Delayed Rail Movements That Could Affect Army Readiness or Deployment Time Lines

If the Army privatizes rail operations at additional installations, it will be relying on the commercial rail carriers to provide additional resources in a timely manner to assist with the loading of tanks and other equipment onto rail cars, as well as movements from these installations to ports or CTC rotations. The resources that are most affected by privatization are rail crews and locomotives, both of which can be scarce resources.[5]

Locomotive engineering is a highly specialized occupation that requires a lengthy training period to become qualified. Installation rail personnel indicated that, because of Army and FRA requirements, it takes two years to become a licensed engineer at an Army installation. Similar training requirements exist for commercial freight rail engineers and other rail crew members. Freight rail companies tend to keep their rail crews occupied and have few in reserve. For example, we were told by MCLB Barstow rail staff that UP has only two rail crews on exclusive reserve for unexpected jobs and emergencies, and if there are unexpected needs during holidays, it may be difficult to find available rail crews. Similarly, locomotives are very expensive assets that commercial rail companies keep in almost continuous use. In contrast, Army locomotives and GOGO or GOCO rail crews can be readily available on short notice.

With sufficient notice, it should be possible for commercial rail carriers to provide rail crews and locomotives for planned movements, such as CTC rotations. However, Army installation rail personnel expressed concerns about whether commercial rail carriers would be responsive, since the Army only accounts for a small fraction of their revenue (less than 1 percent, in some cases). In addition, for short-notice deployments, it could be more problematic to obtain commercial rail crews and locomotives. If rail carriers reallocate them from commercial workloads, they could face penalties for not meeting delivery deadlines.

There is some evidence that installations experience minor delays by relying on commercial rail carriers during peacetime operations. For example, Fort Hood rail personnel stated that BNSF rail crews typically show up 6 to 12 hours late to pick up trains. SDDC provided data on the timeliness of rail carriers in FY 2015, summarized in Table 3.2. Note that just under two-thirds of the loaded rail cars were delivered on or before their original estimated time of arrival (ETA) across all commercial rail carriers. However, the remainder was delivered within five days of the original ETA.

If the commercial rail carriers have difficulties providing rail crews and locomotives for timely line-haul movements, they may also incur delays in providing similar resources to assist with loading and unloading rail cars. It is uncertain how lengthy such delays could be if the Army relies on commercial rail carriers for more services than they currently provide. If the delays are more significant, especially during a time-critical contingency operation, then they have the potential to affect the Army's mission by slowing deployment timeliness or affecting readiness if equipment does not arrive on time for CTC rotations.[6]

---

[5] A third scarce resource is rail cars, but they are not directly affected by privatization. Weisgerber (2013) reports that the U.S. military is facing a shortage of commercially owned chain tie-down flatcars because the existing fleet is reaching the end of its 50-year life-span and must be retired under Federal Railroad Administration (FRA) regulations. DODX cars, which are needed to move the Army's heaviest equipment, are also scarce and must be carefully managed to meet peacetime demand. To address rail car shortages, the Army procured chains that can be used with general-purpose commercial flatcars equipped with holes for anchoring tie-down chains and positioned the chains at major deployment installations in 2015 (SDDC, undated, circa 2014).

[6] The responsiveness of rail carriers may also depend on how well the installation's rail operations fit into their local business model. For example, SDDC TEA personnel noted that some installations are not located near commercial rail yards, so carriers are less likely to consider on-post switching to be compatible with their local business operations.

Table 3.2
Timeliness of Rail Carriers in Fiscal Year 2015

Rail Carrier	Total Cars Delivered	Percent Delivered On or Before Original ETA	Percent Delivered Within Two Days of Original ETA	Percent Delivered Within Four Days of Original ETA
BNSF	4,265	73	90	96
CSX	2,753	61	80	84
KCS	2,327	66	90	94
NS	2,623	62	85	90
UP	9,083	63	86	93
WSOR	68	47	47	76
Other	1,689	64	80	86
Total	**22,808**	**65**	**86**	**92**

SOURCE: SDDC rail shipment data.

NOTES: WSOR = Wisconsin & Southern Railroad; KCS = Kansas City Southern.

Loss of Surge Capacity for Deployments

One of the most significant risks of expanding privatization is that Army installations may lose some of their capacity to surge to meet deployment time lines. Currently, an installation with two GOGO or GOCO rail crews can conduct 24-hour operations for six days before they are required to take a day off under FRA safety regulations.[7] Interviewees also told us that installations within the 407th AFSB's region (shown in Figure 3.4), such as Fort Hood, Fort Sill, and Fort Carson, share GOGO and GOCO crews when loading unit equipment for CTC rotations or deployments, as this allows them to conduct 24-hour operations for a longer period and load an entire BCT's equipment more quickly. In addition, although Fort Carson is a GOGO installation, interviewees told us that, over time, it has lost its authorizations for rail personnel and has only one rail operator. Thus, Fort Carson has to borrow rail crews from other installations or rely on commercial rail carriers whenever it needs to conduct a rail movement. As a result, privatizing rail operations at some installations in the 407th AFSB region could reduce the Army's surge capacity on other installations where it still has GOGO or GOCO crews, unless commercial rail crews are available to assist at these installations.[8] As the number of GOGO and GOCO rail crews is reduced within a region, the Army might need to have access to multiple commercial rail crews at the same time to deploy one or more BCTs.

Similarly, although privatization will not directly reduce rail infrastructure, budget pressures may push the Army in that direction. If rail infrastructure is reduced, such as by removing an extra rail yard, storage track, or wye, the Army could lose some surge capacity. For

---

[7] We should note that, according to Fort Hood rail staff, rail crews sometimes worked ten days straight to get the trains out on time during deployments for Operation Iraqi Freedom.

[8] Interviewees also told us that the last time there were short-notice deployments, for Operation Iraqi Freedom, the Army was a division-based force, so rail personnel could be concentrated at one installation to load its equipment. Since the Army is now a BCT-based force, rail operations may need to surge at multiple installations to meet deployment time lines for some planning scenarios.

Figure 3.4
Installations in the 407th Army Field Support Brigade Region

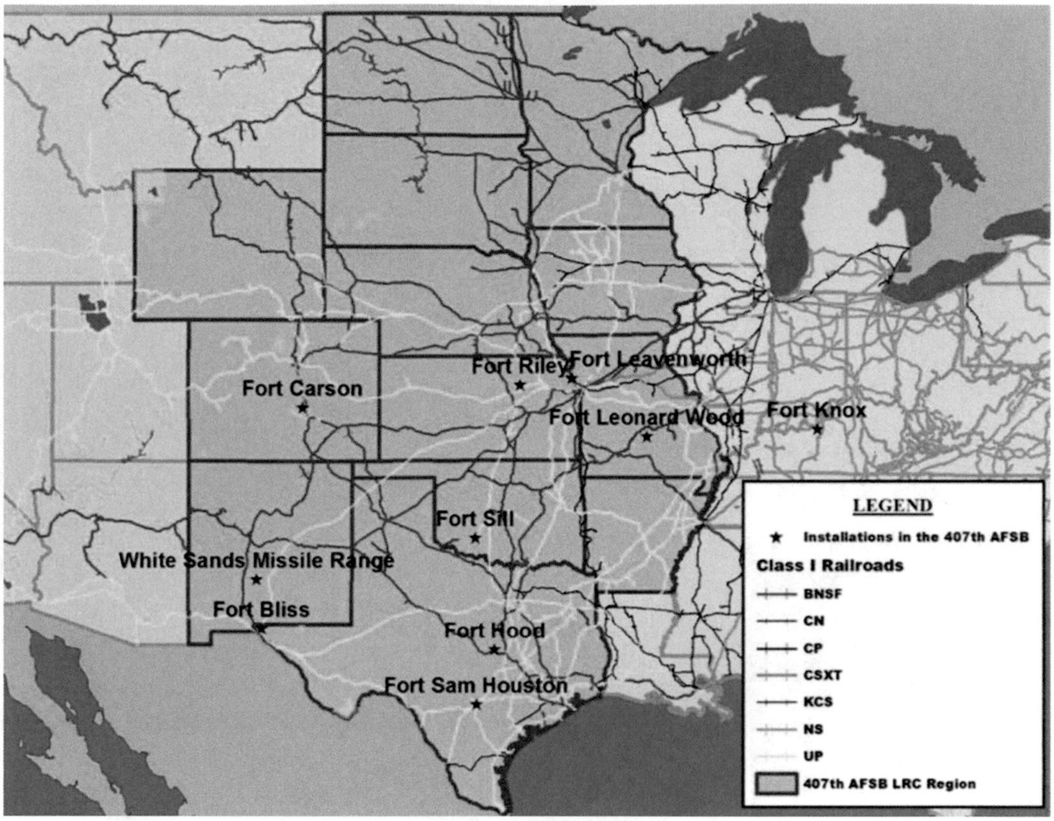

SOURCE: RAND-created GIS map.
NOTES: CN = Canadian National; CP = Canadian Pacific.
RAND RR2009A-3.4

example, Fort Hood personnel told us that a decision had been made in the 1990s to take out its east rail yard, including a second wye, to put in military family housing and other facilities. The installation had just constructed a new rail yard and was moving industrial operations away from its main cantonment area. The older rail yard is not used during peacetime operations; the installation is facing significant facility maintenance budget shortages and does not want to spend money on infrastructure that is not being used and does not seem to be needed. However, since the track had not yet been removed in the early 2000s, the second rail yard's wye and extra track were used to store and turn rail cars during the Operation Iraqi Freedom surge. If removed, the installation would lose the ability to store about 300 rail cars, but they could be stored nearby off post. Without this east rail yard, Fort Hood should still be able to meet deployment requirements in the future, but it would lose some rail operational flexibility and convenience. In contrast, Fort Benning was making efforts to retain its rail and other infrastructure, which would allow it to house a BCT in the future, if needed. Based on examples from contracts and memoranda of agreement (MOAs) at other installations, discussed in Appendix B, there may be opportunities to lease some underutilized Army rail facilities to rail carriers or rail services companies to defray some of the maintenance costs for this infrastructure.

Increased Likelihood of Accidents or Violations of Safety and Environmental Rules

There is a risk of more accidents or violations of safety and environmental rules if the Army relies on commercial rail crews that are unfamiliar with an installation's infrastructure. For example, Fort Sill rail personnel told us that they had a large amount of rail activity in 2001–2002 to support operations, and the Army hired an extra crew from a commercial rail carrier that was not familiar with the installation. This commercial rail crew did some things that were unsafe, increasing the risk of accidents. When rail accidents and safety violations occur, they can potentially cause delays in rail operations, injuries, and loss of life or property. They also can increase the costs of operations. We illustrate with three examples, two from Fort Hood and one from Fort Carson.

In 2005 at Fort Hood, a string of 22 flatbed rail cars carrying M1A2 tanks for the 4th Infantry Division were awaiting movement to the Port of Beaumont, Texas, for deployment to Kuwait. The brakes failed, and the rail cars rolled down a 2 percent grade and collided with five BNSF locomotives idling on the siding near the main line. The derailment damaged 11 tanks with a 1991 purchase price of $3.1 million each. In addition, there was a total loss of three of the BNSF locomotives and five of the DODX rail cars, and 300 to 400 feet of rail siding was also destroyed (Baker, 2005, and Clark, 2005). The accident report found that the rail crew improperly applied the hand brakes and did not use standard external safety devices, such as chock blocks and derails. It also recommended that Fort Hood "develop local rail yard procedures to include an agreement with the commercial rail carrier (BNSF) on when and how BNSF rail carriers can enter the installation, remove/place rail cars, control/position switches, etc." (Garst, 2005). See Figure 3.5 for a photo of some of the damage from this 2005 accident.

Figure 3.5
Damage from the 2005 Fort Hood Accident

SOURCE: Photo by the 407th AFSB.

RAND RR2009A-3.5

Figure 3.6
Damage from the 2011 Fort Carson Derailment

RAND *RR2009A-3.6*

In 2010, an RC brakeman mobilized to support rail operations at Fort Hood failed to set the hand brake, chock the wheel of the car, and verify that the car would not move before uncoupling it. In addition, the RC conductor failed to take appropriate actions to stop violations of safe rail operations because he was not in a physical location to supervise the train movement and ensure that the brakeman followed the proper procedures. As a result, a rail car carrying two M1A1 tanks derailed. The rail car, M1A1 tanks, and track sustained "significant damage" (U.S. Army Combat Readiness/Safety Center, 2010).

In November 2011, Fort Carson also had a derailment. A train with 48 loaded flatcars being pulled by a UP rail crew damaged a switch, causing three rail cars to derail. The switch and track section were down for six months until the repairs were completed. The total repair cost was $289,278, including $111,000 to repair the track, $160,000 to repair the rail cars, and $18,000 for the accrued finance charges while the involved parties were disputing responsibility for payment. The rail crew's unfamiliarity with the installation's rail infrastructure and the location of switches likely contributed to this accident. Figure 3.6 shows two photographs of the damage at Fort Carson.

To some extent, the risk of unfamiliarity with installation infrastructure can be reduced by requiring the commercial rail carriers to provide the same crews on a consistent basis. However, such a policy could increase delays if the customary crew is occupied with a commercial movement when a short-notice deployment occurs. Complacency of GOGO, GOCO, or commercial crews can also be a factor contributing to accidents. A more comprehensive review of rail accidents on Army installations would be needed to assess the extent to which lack of familiarity with rail infrastructure and other factors increase the risk of accidents.

It is important to note that such accidents and other damage to rail infrastructure, caused by natural disasters or terrorist incidents, are current risks that Army installation rail operations already face, regardless of the business model. The incident (discussed in Chapter Two) in which the track washed out due to flooding at Fort Sill in June 2016 provides an example of such a risk. The Army has historically invested in rail infrastructure to help hedge against such incidents. Having diversity, redundancy, and robustness in the Army's rail infrastructure across U.S. installations helps ensure that the Army can meet uncertain future contingency requirements. Reductions in Army rail crews and infrastructure, whether through privatization or taking out track at an installation, could reduce the surge capacity in the Army's rail system and make it more vulnerable to risks.

Unexpected Additional Costs

Another risk of privatization is higher costs during deployments. As noted in the cost analysis, the estimated cost of additional services at installations with privatized rail operations, including switching rail cars and overseeing equipment loading by unit personnel, is in the range of $400 to $900 per rail car. If the number of GOGO and GOCO rail crews is reduced Army-wide, costs could increase at other nonprivatized installations due to a loss of surge capacity and the ability to share rail crews across installations. These installations would then need to pay for commercial rail crews to assist GOGO or GOCO rail crews with switching rail cars and loading equipment during surge operations.

The Army currently contracts separately for each rail movement through SDDC. Representatives of commercial rail carriers said that they liked the flexibility of this system, because they can adjust their bids depending on how busy they are—i.e., charging a higher price when they have plenty of commercial business or offering a lower price when they face competition or other business is slack. However, this short-term contracting approach could work against the Army if it allows the rail carriers to raise their prices when they are most needed by the Army for a short-notice deployment. This risk is higher when an installation is served by a local monopoly rail carrier, because the only alternative may be to use trucks instead of rail.

Rail carrier representatives said that they would give priority to the military during national defense emergencies and time of war; however, they may need to charge more to be able to acquire and supply the needed rail assets. For example, if rail crews and locomotives have to be diverted from commercial movements, the rail carriers may face penalties for failing to meet delivery deadlines.[9]

To reduce the risk of price fluctuations, the Army may need to enter into long-term contracts with rail carriers at installations with privatized rail operations. These contracts should specify prices for movements and additional services in advance, so that the Army will not face unexpected price increases after it becomes more dependent on commercial rail carriers or when rail carriers need to respond quickly for a short-notice deployment.[10]

Additional Risks from Increased Privatization

Army installation rail personnel that we interviewed also raised some additional issues that would need to be addressed in the event of increased privatization of installation rail operations.

First, commercial rail crews might need to pass DoD background checks to work on an Army installation. GOGO and GOCO rail crews on Army installations and at the MCLB rail yard in Yermo deliver and pick up trains near the installation boundary, so commercial rail crews do not need DoD background checks. Requiring background checks might help ensure

---

[9] Another potential problem is moving empty rail cars to installations where they are needed for a deployment. If rail carriers have sufficient notice, they can move empty cars at low cost by adding them to regularly scheduled trains. However, if they need to be moved long distances on short notice, these additional costs could be added to a bid for Army rail services. Rail carrier representatives said that they do not currently charge separate fees for moving empty rail cars.

[10] Contracts for privatized operations should also specify whether the carrier has the right to store commercial rail cars on the installation when not in use. If such storage is allowed, it must be compatible with the installation's contingency load-out operations and not interfere with meeting deployment time lines.

that commercial rail crews are familiar with the installation infrastructure, but it could also limit the availability of crews that could be used to meet a short-term deployment requirement.

Second, privatization may require infrastructure investments to accommodate commercial locomotives or to meet FRA safety requirements. Interviewees told us that most commercial rail carriers use the GE Dash 9 locomotives, which require 115-pound rails, while most Army installations have 100- to 110-pound rails. These commercial locomotives do not perform as well on the 100- to 110-pound rails at most Army installations, and there could be an increased risk of rail car derailments if the Army does not upgrade its tracks. Another issue is that most Class I locomotives have six axles and thus are not compatible with sharply curved installation track, even if it is well maintained. Most Army locomotives have four axles and are more compatible with installation track. In addition, interviewees said that some rail crossings may need to be modified to meet FRA standards.

A third concern is potential contracting challenges and increased regulatory attention. Rail unions, such as the United Transportation Union and Brotherhood of Locomotive Engineers and Trainmen, are powerful, and they would likely have to agree with the contract terms between the rail carriers and the Army. For example, the unions may insist that installations meet FRA safety requirements from which they are currently exempt, such as the requirement for crossing gates at every railroad crossing.[11] Personnel at SDDC TEA also noted that an increase in commercial locomotives entering installations could increase FRA interest in the condition of the Army's track and other rail infrastructure. Either the FRA or the commercial rail carrier would have the right to close rail operations at any Army installation where they felt that the track conditions were unsafe.[12]

Fourth, privatization of rail operations is likely to result in the loss of some Army rail expertise, such as the capability to train unit personnel to load and tie down equipment. Army personnel must follow specialized instructions to safely tie down military equipment onto chained DODX flatcars as specified by SDDC TEA Modal Instruction 55-19 (SDDC TEA, 2015). This expertise may not be readily available at commercial rail carriers.

Comparing Cost Savings and Risks from Privatization

As we noted earlier, our cost analysis identified five installations with deployable units where shipments of loaded rail cars averaged fewer than 30 per month in FY 2015, and these could be candidates for privatization based on possible cost savings. For these installations, the Army should weigh the potential cost savings against possible risks of privatization. Table 3.3 summarizes information about current rail operations and risks at these installations.

---

[11] For example, Ellig (2002, p. 163) notes that three laws give railroad unions greater control over work conditions and higher benefits than in other industries. "The Railway Labor Act essentially gives labor unions a veto over the removal of costly work rules and permits striking railroad employees to picket shippers and other enterprises doing business with railroads. The Federal Employers Liability Act raises the compensation to injured workers by 40–60% above the cost of the regular workers' compensation program employed in other industries. The Railroad Retirement Act costs railroads three times the percentage of payroll that other industries pay for retirement plans. In 1991, the U.S. General Accounting Office estimated that these three railroad labor policies raised the railroad industry's costs by $3 billion annually."

[12] In addition, SDDC TEA personnel said that privatization could increase the likelihood that rail carriers bring commercial commodities onto installations for short periods of time while picking up small numbers of rail cars from the installation (i.e., when there are not enough cars for a dedicated train). If there is any damage to commercial commodities while on the installation, the Army could be partially liable.

Table 3.3
Characteristics of Possible Candidates for Privatization

Installation	Business Model	Average Rail Cars Shipped per Month (FY15)	Rail Deployment Requirement	Other Deployment Considerations	Rail Carrier(s): Local Rail Monopoly Concern?
Fort Bragg	GOCO	24	IBCT-A	PPP 3 IBCT-A, 1 CAB, 1 SB, 1 ADA brigade	CSX: Yes
Fort Leonard Wood	GOGO	<1	Engineer battalion	Maneuver Support Center of Excellence	BNSF: Yes
Fort McCoy	GOCO	3	Engineer battalion	PPP Training, mobilization, and deployment center for ARNG and USAR	CP, UP: No
Fort Sill	GOCO	15	Fires brigade	PPP ADA brigades	SLWC, BNSF, UP: No

SOURCES: U.S. Army Audit Agency, 2017, and SDDC rail shipment data.

NOTES: ARNG = Army National Guard; USAR = U.S. Army Reserve.

The information in Table 3.3 includes the installation's current business model for rail operations; the average number of loaded rail cars shipped per month in FY 2015; the installation's rail deployment requirement; other deployment considerations, including other units stationed at the installation and whether it is a power projection platform (PPP);[13] and rail carriers serving the installation, including whether there is a local monopoly concern. For example, Fort Bragg has GOCO rail operations performed by Cape Fear Railways and it shipped an average of 24 rail cars per month in FY 2015, primarily to the JRTC at Fort Polk. Its deployment requirement is based on shipping the vehicles and equipment needed for combat operations by an airborne IBCT (IBCT-A). Fort Bragg is one of the Army's 15 installations designated as a PPP. Other deployable units include the 82nd Airborne Division's headquarters, two additional IBCT-As, division artillery, combat aviation brigade (CAB), and sustainment brigade (SB), as well as an air defense artillery (ADA) brigade, military police brigade, and fires brigade (U.S. Army Fort Bragg, undated, and DoD Housing Network, 2016). Fort Bragg is served by a local monopoly rail carrier, CSX.

In the remainder of this section, we will discuss the key risk factors for privatization in more detail. These risk factors include the following:

- Does the installation have BCTs or other units that need to meet deployment time lines?
- Does the installation share rail crews with other installations in its AFSB region?
- Is the installation served by only one commercial rail carrier?
- Will infrastructure investments be needed for privatization?

---

[13] A PPP is defined as an Army installation that strategically deploys one or more high-priority active-component brigades or larger units or mobilizes and deploys high-priority RC units. Each has a designated seaport of embarkation and aerial port of embarkation. See, for example, Federal Highway Administration, 2014.

Deployment Requirements

The Army uses rail to move large amounts of heavy equipment over long distances because it can be much less expensive than line-haul trucks. In addition, some types of heavy equipment, such as tanks, are difficult (if not impossible) to move long distances by road because of their weight and dimensions. Therefore, ensuring efficient and effective rail operations is most important at Army installations with ABCTs and those that are a long distance from a port. Table 3.4 shows the locations of BCTs at several U.S. installations, along with their average monthly rail car movements. Note that larger numbers of monthly rail car movements tend to be associated with installations that have heavy BCTs and those that are located farther from ports. Installations with lighter BCTs, such as Fort Bragg and Fort Drum, tend to have less rail activity. All of the installations listed in Table 3.4 are PPPs and have associated deployment requirements.

Sharing Rail Crews

Army rail personnel at Fort Hood and Fort Sill noted that installations frequently share GOGO and GOCO rail crews to assist with preparation of major shipments to CTCs or ports. Thus, any decisions to privatize installation rail operations should consider possible effects on other installations in the same AFSB region and across the United States. Figure 3.7 shows a map of the three U.S. AFSB regions and the installations associated with each.

It is important for the Army to have a robust system, including access to a sufficient number of rail crews, because the Army may need to deploy units simultaneously from multiple locations across the United States. For instance, an operational scenario might require the Army to deploy an ABCT from Fort Hood, an IBCT from Fort Carson, a fires brigade from Fort Sill, and a division headquarters unit from a fourth installation. Given the need for at least three rail crews to sustain 24-hour operations for more than six days, the 407th AFSB

Table 3.4
Number of Brigade Combat Teams at Selected Army Installations

Installation	IBCTs	IBCT-As	ABCTs	SBCTs	Total BCTs	FY15 Average Monthly Rail Car Movements
Fort Hood	0	0	3	1	4	168
Fort Carson	1	0	1	1	3	108
Fort Riley	0	0	2	0	2	67
Joint Base Lewis-McChord	0	0	0	2	2	68
Fort Stewart	1	0	1	0	2	52
Fort Campbell	3	0	0	0	3	42
Fort Bliss	0	0	2	1	3	40
Fort Drum	2	0	0	0	2	34
Fort Bragg	0	3	0	0	3	24

SOURCE: Stoneburg and Lyle, 2015.
NOTE: SBCT = Stryker Brigade Combat Team.

Figure 3.7
Map of U.S. Army Field Support Brigade Regions and Installations

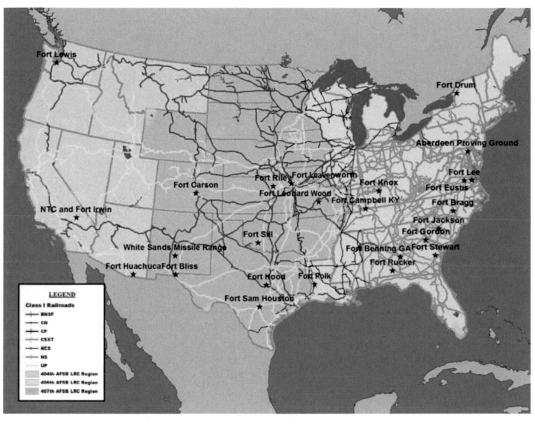

SOURCE: RAND-created GIS map.
RAND RR2009A-3.7

may need to borrow one or more rail crews from another AFSB region to support such a contingency, as it has sometimes done in the past, or hire commercial rail crews.

If additional GOGO or GOCO rail crews are not available to assist, costs could increase at installations that are not privatized if they have to purchase those services from commercial rail carriers. Thus, the Army may want to optimize the number and location of rail crews and locomotives by region, rather than just considering costs at individual installations.

Local Monopoly Concerns

A third consideration is whether the installation is served by only one commercial rail carrier or multiple carriers. After a privatization decision has been made, the Army will be dependent on the rail carriers that serve that installation to provide locomotives and rail crews at a reasonable cost when needed. Local monopoly rail carriers are likely to charge higher prices (relative to their actual costs) and also have less incentive to be responsive when they are not subject to competition from other carriers.[14] Of the 16 installations examined for this study, 6 were served by

---

[14] Entry of new commercial rail carriers is unlikely due to the high capital costs of building and maintaining track. Since deregulation of the rail industry in 1980, the number of Class I railroads fell from 40 in 1980 to 7 in 1999, and the total size of the rail network controlled by those carriers dropped from 164,822 miles in 1980 to 95,391 miles in 2013. See McKenzie, 2016.

more than one rail carrier (Forts Carson, Irwin, Knox, McCoy, and Sill and Joint Base Lewis-McChord), and the remainder had a local monopoly rail carrier.

Even if the installation is served by more than one carrier, a long-term contract with one carrier to provide on-post switching could reduce competition for line-haul service or create friction between separate providers of each service.

Infrastructure Investments

A need to make infrastructure investments to accommodate commercial rail carriers could offset some of the potential cost savings of privatization. For example, as discussed previously, the Army may need to increase the weight of some of its tracks or upgrade on-post railroad crossings. It will be important to find out from rail carriers what investments would be needed and to include their costs in any installation-specific business case analysis for privatization.

Mitigating Risks from Privatization

Some of these risks can be mitigated by specifying contract terms and conditions with rail carriers before privatizing installation rail operations. Long-term contracts with rail carriers should specify the cost and availability of rail crews and locomotives for peacetime and surge operations, including any contingency plans for leasing locomotives or providing additional rail crews for short-notice deployments and the Army's priority relative to other customers. They should also specify requirements for rail crews, such as familiarity with installation infrastructure, DoD background checks, and adherence to safety, security, environmental, and other rules.[15] In addition, contracts should specify the cost of services such as switching cars and overseeing unit personnel during equipment loading and unloading; time lines for notification and performance of loading and unloading operations and deliveries in peacetime and surge operations; and any incentives or penalties for meeting or failing to meet contract terms and conditions. Rail carriers should also be required to notify the Army of any infrastructure improvements needed to accommodate their equipment, FRA requirements, or union agreements.

The competition for the Fort Sill LRC contract that is planned in FY 2017 offers the Army an opportunity to compare the costs and risks of privatization. As part of the LRC contract competition, bidders should be required to separately specify the costs associated with GOCO rail operations. These costs can then be compared with a separate solicitation of bids from rail carriers to provide additional services if the Army decides to privatize rail operations. As discussed earlier, these bids should include the cost and availability of rail crews and locomotives for peacetime operations and short-notice deployments; prenegotiated rates (or cost-based formulas) for rail movements, including additional services, such as switching rail cars and overseeing loading and unloading by Army unit personnel; and any infrastructure investments needed. In addition to making a full comparison of costs between the GOCO and privatization proposals, the Army should consult with the 407th AFSB regarding potential regional effects of privatization on the Army's ability to share rail crews across installations.

---

[15] Based on our review of existing rail contracts and MOAs, there are precedents for many of these types of contract terms and conditions. However, such specifications could increase the price of services and erode the potential cost savings from privatization.

CHAPTER FOUR

Conclusions

Based on our analysis, the Army may be able to obtain some modest cost savings by privatizing rail operations at installations with low rail activity rates that currently have GOGO or GOCO business models.[1] These savings would most likely be less than $300,000 per year at each installation. However, the Army should carefully balance these estimated cost savings with the risks of increasing reliance on commercial rail carriers. The most important consideration is whether commercial rail carriers will be able to provide rail crews and locomotives to meet the Army's time lines for short-notice deployments. Second, there may be some loss of internal surge capacity, because installations currently share GOGO and GOCO rail crews in order to staff 24-hour operations when loading a BCT's equipment for a shipment to a port or CTC.[2] Third, there may be an increased likelihood of accidents or violations of safety and environmental rules if commercial carriers send crews that are not familiar with the installation's rail infrastructure. Finally, there may be a risk of unexpected costs, because commercial rail carriers currently bid separately for each shipment. Some rail carriers may have local monopoly power to raise prices unless the Army enters into longer-term contracts that specify prices in advance. Another cost-related consideration is whether additional investments will be needed to accommodate heavier commercial locomotives or to meet FRA standards.

When making privatization decisions at specific installations, the Army should consider factors such as the installation's deployment requirements, whether it shares its rail crews with other installations, whether it is served by a local monopoly rail carrier, and the cost of any infrastructure investments that would be needed to support privatization. Some of the risks of privatization can potentially be mitigated by specifying contract terms and conditions with rail carriers before privatizing installation rail operations. For example, these contracts should specify the cost and availability of rail crews and locomotives for peacetime and surge operations, including contingency plans for short-notice deployments. They should also specify any requirements for rail crews, such as familiarity with the installation rail infrastructure, DoD background checks, and adherence to safety, security, and environmental rules.

The competition for the Fort Sill LRC contract that is planned in FY 2017 offers the Army an opportunity to compare the costs and risks of privatization. As part of this competition, the Army should require bidders to separately specify the costs associated with GOCO rail operations so that they can be compared with bids from rail carriers regarding the costs of additional services they would provide if the Army decides to privatize.

---

[1] These installations include Forts Bragg, Leonard Wood, McCoy, and Sill.

[2] If commercial carriers are able to provide additional crews to meet surge demands, nonprivatized installations would incur any additional costs of these services.

Marine Corps Logistics Base Barstow Yermo Annex Rail Yard Support for Fort Irwin

This appendix describes rail operations for units training at the NTC. Since there are no rail lines onto Fort Irwin, the trains are loaded and unloaded at MCLB Barstow's Yermo Annex rail yard. Marine Corps personnel at Yermo told us that an ABCT typically has eight trainloads of equipment that arrive each month over a period of several days, with a total of 400 to 430 rail cars (approximately 60 rail cars per train, based on rail carrier limits). It takes about seven days to unload the equipment and another seven days to load it after the training rotation is completed.

From the Yermo rail yard, all the unit equipment that comes off the trains has to travel the 36 miles to Fort Irwin. The wheeled vehicles are driven by convoy on the dirt Mannex trail, and the tracked vehicles (tanks and engineering equipment) and containers are transported by intermodal truck service to Fort Irwin. Each convoy consists of 40 to 50 vehicles, with about 14 to 16 convoys required per unit. It takes four to five days for all of a unit's convoys to make the trip. Intermodal truck service is provided by two commercial line-haul truck companies and Army Heavy Equipment Transporters. The trucks are loaded and unloaded at the Yermo rail yard and Fort Irwin's Dust Bowl (see Figure A.1).

The MCLB Barstow Yermo Annex rail yard is the largest military train depot in the continental United States, with 80 percent of its workload supporting the NTC. In 2013, this rail yard processed more than 50 million pounds of equipment and vehicle freight transfer every month. In 2015, it handled 60 percent of all DoD rail traffic, including that of the Army, Marine Corps, Navy, Air Force, and Coast Guard (Beckstrom, 2014). The Yermo Annex rail yard is serviced by two major rail carriers, just off the base: BNSF, whose rail line ends seven miles away at Daggett, California, and UP, whose line goes to the edge of Yermo. On Yermo rail yard, two Army-owned engines are operated as pairs.

At the time of our visit in August 2016, the rail yard was run by a retired Marine who is now a Marine Corps civilian employee. Rail yard employees include 10 DoD civilians, 15 contractor employees, and 21 Army soldiers from the Army 916th SB Rail Transportation Unit rail detail. The rail detail helps train and supervise the soldiers from arriving and departing units as they load and tie down or untie and unload their equipment. During rail operations, some soldiers and DoD civilians stay overnight at the rail yard.

Figure A.1
Trucks Being Unloaded at Fort Irwin's Dust Bowl

SOURCE: Photo by Beth Lachman.
RAND RR2009A-A.1

Installation Railroad Contracts and Agreements

As part of this study, we also examined Army installation contracts, memoranda of agreement, and other types of agreements for rail-related services. This appendix summarizes our review of these documents. Examining the terms and language used in such contracts and agreements is useful for understanding some of the opportunities, as well as challenges, that would be involved if Army installations enter into more partnerships with rail companies and other organizations that can help provide rail movement or infrastructure services. In addition to contracts and agreements for rail-related services, we provide some examples in which Army installations have shared, leased, or sold some rail assets for fees and in-kind services, such as rail infrastructure maintenance. Although many of these agreements involve industrial facilities, such as depots, arsenals, and ammunition plants, they may also offer opportunities for installations with deployable units to reduce the costs of maintaining infrastructure that is used infrequently for deployments.

Range of Functions and Types of Army Installation Rail Contracts and Agreements

Depending on each installation's history, needs and use of rail, and business model for on-post rail operations, there are a range of rail-related services that may be provided by the contractor or partner as specified in the contracts and agreements. These services can include the partner or contractor providing rail movement operations, a range of installation transportation rail services, or repair and maintenance of the installation's railroad system both on and off post; operating and maintaining Army locomotives; using track on the Army installation for the movement and storage of rail cars (typically paying a fee for usage); sharing use of the rail lines with the Army installation; and leasing installation property when performing rail car maintenance. We describe and provide examples of each of these types of agreements here.

In some cases, the rail contract is included in the LRC Performance Work Statement (PWS), where rail services are specified as a small part of a larger LRC contract at installations with GOCO or privatized rail operations, such as at Forts Benning, Bliss, Campbell, McCoy, and Polk. In other cases, the installation may have a separate contract for installation transportation rail services (ITRS) in accordance with the PWS, such as at Fort Knox. In addition, some Army installations have separate contracts for installation railroad system repair and maintenance, such as at Fort Campbell. We also found at least one example of a contract for use of an installation's rail system. This contract was at Pine Bluff Arsenal, where Lindsey and Osborne Pine Bluff, LLC, pays to use 15 miles of track for the movement and storage of clean, empty rail cars. Under this contract, the company pays an estimated $10,250 per month, or about $123,000 per year over a five-year period (Contract Number W52P1J-13-C-DF01, 2013). This example illustrates that the Army can, in some cases, lease some of its rail infrastructure to earn funds when the Army is not using the rail assets.[1] Personnel at SDDC TEA noted that BNSF has expressed an interest in storing rail cars on some military installations, and other carriers may also be interested in similar arrangements, which could help defray the costs of maintaining installation rail infrastructure.

Some installations have also established MOAs and other types of agreements with contractors and other organizations regarding installation rail services and infrastructure. These types of agreements include the use of installation track in exchange for rail services and maintenance, railroad use agreements, and tenant use agreements. We provide three different installation examples. Fort Bragg had an agreement with the Cape Fear Railways for use of tracks at Fort Bragg in exchange for providing installation utility rail services and rail system maintenance services (Fort Bragg, 1994).[2] Holston Army Ammunition Plant in Kingsport, Tennessee, had a tenant use agreement under which the tenant, Appalachian Railcar Services, Inc., used two buildings and adjacent yard areas on the installation "for the purposes of Railcar Maintenance, Repair and Painting and Coating Operations." The tenant paid the Army $150,000 per year as part of this leasing arrangement.[3] The third example is a unique agreement at Letterkenny Army Depot (LEAD) in Pennsylvania to transfer part of LEAD's rail infrastructure as part of a 1990 base realignment decision. Using an MOA, the Army conveyed (through a sale) most of the installation rail lines to Letterkenny Industrial Development Authority (LIDA), while the Army retained some rail equipment and facilities, such as three locomotives and an engine house. The partners share use and cost of the rail lines and facilities, and the Army has priority use during mobilization (Department of the Army, 1998a and 1998b).

Common Elements of Installation Rail Contracts and Agreements

We found that Army installation rail contracts and agreements tended to have some common elements. We have grouped these common elements into two main areas: statements about rail services provided and who owns and uses the rail assets, and key terms and conditions. The items in each category are as follows:

- statements about the provision of rail services and rail asset ownership and use
  - installation transportation rail services to be provided by the contractor or partner to the Army installation
  - who maintains and upgrades selected installation rail assets and who pays for them
  - who owns which rail assets
  - who uses which rail assets

---

[1]   Since the Army still owns its rail infrastructure (e.g., track, storage yards, and loading ramps) under the privatized business model, decisions to lease underutilized assets are currently independent of privatization decisions and would not directly affect cost comparisons.

[2]   This agreement excludes Fort Bragg track located near the Ammunition Supply Point.

[3]   The lease ran from November 1, 2006, through October 31, 2009 ("Tenant Use Agreement, Holston," 2006, p. 1). This agreement was later amended and extended. For example, an amended lease agreement runs from January 1, 2013, through December 31, 2018, with a rent of $172,000 per year ("Amendment," 2013).

- key terms and conditions regarding the rail contract or agreement
  - Army rail priority for mobilizations and other national interest situations
  - security requirements
  - other rules and regulations that the contractor or partner must follow
  - liability issues
  - accountability statements.

We describe each of these areas in the following sections.

Provision of Rail Services and Rail Asset Ownership and Use

In this section we describe the range of contract and agreement statements in each of the four subareas of this category and illustrate them with some examples.

Installation Transportation Rail Services

Most of the installations that employ the GOCO or privatized business model have contracts specifying the types of rail services to be provided to the installation. For instance, at Fort Campbell, which has GOCO rail operations, the LRC PWS states, "The Contractor shall conduct all portions of rail operations in accordance with DoD 4500.9-R and AR 56-3." This document goes on to state that the contractor shall operate three locomotive engines and perform specific functions that "include performing braker/switcher duties, ordering fuel and oil for locomotives, protecting all rail crossings between all points on and off the installation on the Fort Campbell Rail System, and performing interchange operations with commercial rail carriers" (Fort Campbell, 2012, p. 53). As this Fort Campbell example illustrates, the contracts specify the conditions under which the contractor or partner provides rail cars, rail planning, switching services, inspection, and other installation operational rail services. Similarly, at Fort Benning (an example of the privatized model) the LRC PWS also states, "The Contractor shall conduct all portions of rail operations in accordance with DoD 4500.9-R and AR 56-3." However, in this case, the LRC contractor is primarily responsible for coordination with the commercial rail carrier rather than performing rail operations on the installation. The contract (Fort Benning, 2015, p. 3) specifies, among other things, that the contractor shall

- "assist with the preparation, update, and distribution of the rail loading plan in sufficient time to obtain carrier equipment and meet deployment load out schedules, unit practices, or training exercises"
- "coordinate with the rail inspector for the pre-loading inspection of all rail cars"
- "order required rail service after notification of routing from SDDC"
- "provide support for rail loading and unloading operations"
- "contact the carrier for the switch out of rail cars"
- "assist in providing technical advice for the positioning and securing of equipment on cars; and blocking, bracing, and the tying-down of equipment"
- "perform final inspection of carrier's equipment with the carrier representative."

Some contracts and agreements also include services related to installation rail infrastructure inspections. For example, the Fort Bliss LRC PWS spells out that the contractor should conduct all rail operations, including the grade crossing inventory and rail inspections; inspection of loaded rail cars; and the operation, inspection, and supervision of the Rail Deployment Area and facilities (Fort Bliss, 2014, Section C-5, pp. 58–59).

Maintenance and Upgrades of Selected Installation Rail Assets

Many Army installations also have provisions or separate specialized contracts or agreements for the contractor or partner to maintain or upgrade locomotives, track, and other rail infrastructure. For example, the Fort Bliss LRC PWS states that the contractor shall "perform appropriate field level maintenance of Army owned or leased locomotives and Army owned USA/USAX rail cars . . . to include pre-, during, and post operational checks and services of locomotives, including fueling / re-fueling" (Fort Bliss, 2014, Section C-5, p. 58). Similarly, at Fort Campbell the LRC PWS specifies that the contractor shall maintain three locomotives. At Fort Sill the LRC PWS specifically mentions rail equipment repair, stating that "the contractor shall perform daily operator maintenance and quarterly organizational maintenance and repair of railway equipment (rolling stock)."[4] At Fort Bragg the rail partner is required to provide a range of installation rail maintenance functions, including maintaining, repairing, protecting, and preserving the installation government-furnished property; maintaining 16.06 miles of track (including spurs and sidings) in Class 2 conditions in accordance with FRA track safety standards; and maintaining and keeping operational at all times all electrically operated warning devices and cross-arm signals (Fort Bragg, 1994).

The documents also specify who pays for these services. The contractor or partner may provide maintenance and upgrades on the installation rail system for a fee from the Army or in exchange for using the track and other parts of the rail system. For example, in the Fort Bragg agreement the rail partner maintains the installation track and other equipment in exchange for using the track (Fort Bragg, 1994). At Letterkenny Army Depot, LIDA provides the track maintenance based on LEAD's requirements. The Army pays a track maintenance fee to LIDA for the portion of the track that it uses on a per-mile basis, "to defray costs incurred as a result of Army requirements to maintain track in readiness-to-serve condition" (Department of the Army, 1998b, p. 3).

Ownership of Rail Assets

Some of the rail documents also clarify who owns what installation rail assets, especially if there is transfer or sharing of assets. These provisions can include a range of rail assets, such as locomotives, engine houses, track, and switches. For example, the Fort Bragg agreement states that the Army government-owned property and facilities on the installation include 16.06 miles of track, 45 switches, nine electric crossing signals, and two locomotives, as well as a depot, a maintenance shop, and a tool shed. It lists these specific facilities because they are the ones the rail partner is allowed to use. At LEAD, the rail agreement states that the Army retains ownership of three locomotives, an engine house, and related maintenance equipment.

Conveyance of Army rail assets to a partner is a special subset of ownership. Such conveyance can include track, buildings, locomotives, equipment, and facilities. For instance, the LEAD Railroad Use Agreement conveyed installation rail lines to LIDA, except for the rail lines in the ammunition area (Department of the Army, 1998b).

Usage of Rail Assets

The contracts or agreements also may state when and how Army and contractor or partner rail assets may be used by each party. These assets can include rail lines, buildings, locomotives, and other railroad infrastructure. Conditions of use are most often specified when partners are

---

[4] Contract excerpt provided by Fort Sill LRC personnel.

sharing, leasing, or otherwise using installation rail assets. For example, at Fort Bragg, the rail partner may use 16.06 miles of the installation standard gauge of trackage (including spurs and sidings), 45 switches, nine electric crossing signals, and two locomotives. The partner may also use and occupy Army-owned buildings and facilities, including a depot, a maintenance shop, and a tool shed (Fort Bragg, 1994). At LEAD, LIDA may obtain temporary use of LEAD's engine house through coordination with the installation commander.

If there is a fee for usage, that is also specified in the document. For instance, at Pine Bluff Arsenal, the contractor leases and uses 15 miles of track for the movement and storage of clean, empty rail cars and pays the Army about $123,000 per year. The Army also agrees to supply some of the rail cars.[5] Similarly, Holston Army Ammunition Plant has a tenant use agreement under which Appalachian Railcar Services, Inc., pays approximately $172,000 per year for the use of two buildings and adjacent yard areas on the installation "for the purposes of Railcar Maintenance, Repair and Painting and Coating Operations."[6]

Easements for the use of rail lines and facilities are a special case. For example, at LEAD, the Army sold LEAD's rail lines, but it has an easement for the "non-exclusive right to use the rail" (Department of the Army, 1998b, p. 1).

Key Terms and Conditions in Rail Contracts and Agreements

In this section we describe various terms and conditions commonly found in rail contracts and agreements, organized into four subcategories. We illustrate each type with some examples.

Army Rail Priority for Mobilizations and Other National Interest Situations

Some of the documents directly state the Army's priority for rail services during mobilization or other national interest circumstances. These provisions are often included when the partner or contractor shares some rail assets. However, given some of the risks associated with relying on commercial rail companies for services during deployments and other national defense needs (as we discuss in Chapter Three of the main text), this type of clause may be important for other rail service agreements if the Army decides to increase privatization of installation rail operations.

We provide the language used in two different installation examples. The Pine Bluff Arsenal Contract states, "Under unusual circumstances, when the national interest of the United States so requires, the U.S. Government reserves the right to cancel or suspend all or part of its performance of this contract at any time prior to the delivery of the supplies" (Contract Number W52P1J-13-C-DF01, 2013, p. 2). The Letterkenny Railroad Use Agreement states, "The LIDA and its operator will recognize LEAD's priority for rail service during periods of mobilization and as required for the interest of national defense; and will subordinate all other rail demand during mobilization in the interest of national defense" (Department of the Army, 1998b, p. 2).

Security Requirements

Another important issue that is spelled out in most of the contracts and agreements is security requirements to safeguard installation rail operations and working areas. These requirements

---

[5] This contract applies from October 1, 2013, through September 30, 2018. Total estimated cost for the entire five years is $615,000, with estimated payments to the Army of $10,250 per month (Contract Number W52P1J-13-C-DF01, 2013).

[6] This lease agreement runs from January 1, 2013, through December 31, 2018 ("Amendment," 2013).

apply to contractor and partner personnel who come onto the Army installation and operate or access parts of the installation rail system. We provide some examples from three different installations to illustrate the range of security requirements. First, the Fort Knox ITRS contract states a range of security requirements for the contractor, including the following: all contractor personnel must obtain and maintain a favorable National Agency Check and have proper photo identification, must submit to search of locomotives and rail cars by military police, and must safeguard all government equipment, information, and property. An example of the language for the last of these requirements is, "At the close of each work period, government facilities, equipment, and materials shall be secured" (Contract Number W9124D-13-H-0001, 2012, p. 9). Second, the Fort Campbell contract states that the contractor's personnel must complete antiterrorism awareness training annually. This is a relatively unusual requirement that we did not find in many of the documents examined. Third, at Pine Bluff Arsenal, where the contractor uses track on the installation, the contractor's employees must display an approved personnel badge and vehicle permit and must be U.S. citizens; firearms of any kind are prohibited; and the contractor's "personnel will be subject to vehicle search (for contraband, etc.) upon entering/leaving the plant site" (Contract Number W52P1J-13-C-DF01, 2013, p. 6).

Other Rules and Regulations

In addition to security requirements, the contracts and agreements usually spell out other rules and regulations that the contractor or partner and its employees must follow. These requirements include obtaining insurance coverage and adhering to federal, state, and local government safety, environmental, and other rules and regulations, such as those of the FRA, U.S. Environmental Protection Agency, and state environmental agency. Many agreements specify a requirement to follow FRA rules, such as at Fort Campbell, where the LRC PWS states that the contractor "shall operate and maintain three Locomotive Engines . . . in accordance with Federal Railroad Administration (FRA) regulations and guidance."

Sometimes the contract or agreement has language that directly mentions locomotive operators and the need to follow local installation rules. For instance, at LEAD, the agreement requires LIDA to follow the rules and regulations of the post commander and states that shortline and locomotive operators and inspectors must be certified by the appropriate regulatory bodies. Actually, many of the contracts and agreements, especially those that include maintenance tasks, also mention installation-specific rules pertaining to issues such as working hours and conduct for contractor and partner personnel when they are on the Army installation. We illustrate the range of these requirements with the Fort Campbell repair and maintenance contract, which specifies that the contractor must follow the Buy American Act, perform the work during normal work hours, have the necessary excavation and utility clearances, follow conduct and dress requirements (e.g., "Profanity is strictly forbidden"), clean up the work site, and comply with all environmental laws. Similarly, the contract at Pine Bluff Arsenal specifies that the rail contractor shall not construct any structure on the property without Army permission; employees must obey all applicable arsenal safety, security, and traffic rules; and the company must have insurance coverage.

Contracts may also include rules about usage of utilities and other installation services and who supplies and pays for them when the partner or contractor is using installation facilities. For instance, in the Holston Army Ammunition Plant tenant use agreement, the tenant must follow rules and must pay for some installation services, such as water and sewer utilities. Examples of these requirements include the following: "Tenant shall comply with all applicable local, state and federal laws, ordinances, and regulations with regard to construction, sanitation, licenses, and permits to do business, environmental and all other matters," and will pay "a monthly service fee of $150" for the "water, sewer and steam utilities to facilities occupied by Tenant." This document also states that the Army can charge reasonable fees for any services provided, which can include fire protection, security, and emergency response ("Tenant Use Agreement, Holston," 2006, p. 3).

Liability Issues

As with most contracts, the documents often specify liability for property damage, personal injuries, environmental concerns, and other legal issues. We provide three examples to illustrate the type of liability language used. First, the Fort Knox ITRS contract states, "Indemnity. The contractor shall hold the government harmless for any damage to or loss of property, or any injury to or death of persons because of the action or inaction of the contractor or its employees" (Contract Number W9124D-13-H-0001, 2012, p. 13). Second, the contract for the repair and maintenance of Fort Campbell's on-post and off-post rail system states that the contract's provisions are contingent on appropriated funds being available. Specifically, it says, "No legal liability on the part of the Government for any payment may arise for performance under this contract beyond 30 September 2012, until funds are made available to the Contracting Officer for performance and until the Contractor receives notice of availability, to be confirmed in writing by the Contracting Officer" (Solicitation Number W91248-12-R-0010-0004, 2012, p. 60). Third, the Fort Bragg rail agreement states, "Railroad shall be responsible for and held liable for any loss or damage to the rail network and any other Government owned property which results from simple negligence, willful misconduct, or lack of good faith on the part of the Railroad" (Fort Bragg, 1994, p. 3).

Because of concerns about hazardous materials and other environmental issues, the documents sometimes specifically mention environmental liabilities. For example, the Letterkenny Railroad Use Agreement states that "the LIDA's liability, including any environmental liability, shall be as stated in the MOA executed between the parties and the easement documents. . . . In the Army's use of the LIDA owned facilities, the liability of the Army, including any environmental liability, shall be governed by the terms of the MOA executed between the parties; the deeds transferring ownership of the facilities" (Department of the Army, 1998b, p. 4).

Accountability Statements

Finally, most of the documents spell out appropriate contract or agreement metrics and oversight, as well as consequences for not following the terms of the agreement. Some agreements list inspection requirements, especially for maintenance tasks performed under the agreement, and penalties for lack of compliance, such as late fees for not completing a task on time (e.g., track maintenance). Conditions under which the contract or agreement can be terminated may also be specified. For example, at Fort Campbell, the rail repair and maintenance contract describes initial, follow-up, and work-completion inspections and procedures for dealing with deficiencies in the work, including a time line for fixing them. This contract also provides a detailed specification of a late fee for each day the work is past the contract deadline: "If the Contractor fails to complete the work within the time specified in the contract, the Contractor shall pay liquidated damages to the Government in the amount of $219.96 for each calendar day of delay until the work is completed or accepted" (Solicitation Number W91248-12-R-0010-0004, 2012, p. 52).

We illustrate these types of statements with a second example from Pine Bluff Arsenal, where, along with guidelines for how to resolve and prevent disputes, the contract states (Contract Number W52P1J-13-C-DF01, 2013, p. 2),

- "The U.S. Government agrees to promptly notify the Buyer in the event the performance of this contract is canceled or suspended."
- "The Buyer may cancel this contract at any time by providing written notice to the U.S. Government. In this event, the Buyer understands and agrees that it is liable for the costs incurred by the U.S. Government as a result of such cancellation."
- The contractor must provide a standby letter of credit for $25,000 to cover any liability caused by the contractor's "non-performance on this agreement."

Bibliography

"Amendment to the BAE/OSI and Appalachian Railcar Service Facility Use Agreement (Lease) Regarding Railcar Storage, Repair and Maintenance," January 2013.

Assistant Secretary of Defense for Manpower and Reserve Affairs, "Update on OMB Circular A-76 Public-Private Competition Prohibitions—FY 2016," memorandum, April 21, 2016. As of December 1, 2016: http://www.acq.osd.mil/dpap/dars/pgi/docs/Update_on_OMB_Circular_A-76_Public-Private_Competition _Prohibitions-FY2016.pdf

Association of American Railroads, "Class I Railroad Statistics," May 3, 2016. As of April 11, 2017: https://www.aar.org/Documents/Railroad-Statistics.pdf

Bacchus, Jennifer, "DGRC Is Army's Sole Locomotive Overhauler," press release, January 16, 2012. As of November 25, 2016: https://www.army.mil/article/74040

Baker, Emily, "Failed Brake Suspected in Fort Hood Derailment," *Killeen Daily Herald*, November 10, 2005. As of December 7, 2016: http://kdhnews.com/news/failed-brake-suspected-in-fort-hood-derailment/article_b42a9e59-281d-573a-af4a -c61c55121f25.html

Beckstrom, David Nathaniel, "Freight for the Military," press release, September 15, 2014. As of March 22, 2017: https://www.army.mil/article/133734/Freight_for_the_Military

Bureau of Transportation Statistics, "Table 1-57: Tonnage of Top 50 U.S. Water Ports, Ranked by Total Tons," undated. As of December 1, 2016: http://www.rita.dot.gov/bts/sites/rita.dot.gov.bts/files/publications/national_transportation_statistics/html /table_01_57.html

Clark, Harper Scott, "Hood Tanks Damaged in Derailment," *Temple Daily Telegram*, November 9, 2005. As of December 7, 2016: http://www.tdtnews.com/archive/article_07139f92-c1b1-55fc-9040-e8751ccfa323.html

Contract Number W52P1J-13-C-DF01, "Contract of Sale" [at Pine Bluff Arsenal], agreement between Lindsey & Osborne Pine Bluff, LLC, White Hall, Ark., and U.S. Government, November 4, 2013.

Contract Number W9124D-13-H-0001, "Installation Transportation Rail Services (ITRS)," for Fort Knox, October 1, 2012.

Department of Defense Inspector General, *Enhanced Army Global Logistics Enterprise Basic Ordering Agreements and Task Orders Were Properly Executed and Awarded*, Report No. DODIG-2014-095, July 25, 2014. As of November 28, 2016: https://media.defense.gov/2014/Jul/25/2001713385/-1/-1/1/DODIG-2014-095.pdf

Department of the Army, "Memorandum of Agreement Between the Department of the Army and Letterkenny Industrial Development Authority for the Transfer of a Portion of Letterkenny Army Depot, Pennsylvania," November 5, 1998a.

Department of the Army, "Railroad Use Agreement," Exhibit M of the "Memorandum of Agreement Between the Department of the Army and Letterkenny Industrial Development Authority for the Transfer of a Portion of Letterkenny Army Depot, Pennsylvania," November 5, 1998b.

DoD Housing Network, "Fort Bragg, NC Units," 2016. As of December 1, 2016:
http://www.dodhousingnetwork.com/army/fort-bragg/units

Ellig, Jerry, "Railroad Deregulation and Consumer Welfare," *Journal of Regulatory Economics*, Vol. 21, No. 2, March 2002, pp. 143–167.

Federal Highway Administration, "Appendix A. List of Power Projection Platforms," *Coordinating Military Deployments on Roads and Highways: A Guide for State and Local Agencies*, April 1, 2014. As of December 1, 2016:
http://ops.fhwa.dot.gov/publications/fhwahop05029/appendix_a.htm

Fort Benning, "Logistics Readiness Center: Fort Benning, GA, Performance Work Statement," May 2015.

Fort Bliss, "Logistics Readiness Center: Fort Bliss, TX, Performance Work Statement," March 2014.

Fort Bragg, "Rail Agreement for Track Located Within the Boundaries of the Fort Bragg Military Reservation (Excluding the Ammunition Supply Point)," September 22, 1994.

Fort Campbell, "FCKY Logistics Support Contract: Fort Campbell, Kentucky, Performance Work Statement," October 15, 2012.

Fort Polk, "Logistics Readiness Center: Fort Polk, LA, Performance Work Statement," October 17, 2013.

Garst, Douglas L., "Headquarters, III Corps Accident Report Case Number 20051107," memorandum for garrison commander, 2005.

McKenzie, Taylor K., "Technological Change and Productivity in the Rail Industry: A Bayesian Approach," working paper, University of Oregon, November 2016.

National Council for Public-Private Partnerships, "TRADOC Command-Wide A-76 Study of Directorate of Public Works and Directorate of Logistics," 2002. As of March 15, 2017:
http://www.ncppp.org/resources/case-studies/operation-and-managementmaintenance-contracts/tradoc
-command-wide-a-76-study-of-directorate-of-public-works-and-directorate-of-logistics/

Office of Management and Budget, Circular No. A-76 Revised, May 29, 2003. As of December 1, 2016:
https://www.whitehouse.gov/omb/circulars_a076_a76_incl_tech_correction

OMB—*See* Office of Management and Budget.

SDDC—*See* U.S. Military Surface Deployment and Distribution Command.

SDDC TEA—*See* U.S. Military Surface Deployment and Distribution Command, Transportation Engineering Agency.

Solicitation Number W91248-12-R-0010-0004, "Repair and Maintenance of Fort Campbell's On-Post and Off-Post Rail System, Fort Campbell, Kentucky," solicitation, offer, and award, Project Number Fe10217-1J, September 15, 2012.

Stoneburg, John, and Clay Lyle, "Force Structure and Force Design Updates," PowerPoint briefing, August 6, 2015.

"Tenant Use Agreement (Building Lease) for Facilities and Equipment at Holston Army Ammunition Plant, Kingsport Tennessee," November 3, 2006.

"Tenant Use Agreement (Building Lease) for Premises and Equipment at Radford Army Ammunition Plant, Radford, Virginia," 2013.

Tomkins, Richard, "Army Support Task Order Given to VS2," UPI, May 22, 2015. As of November 28, 2016:
http://www.upi.com/Business_News/Security-Industry/2015/05/22/Army-support-task-order-given-to
-VS2/2971432261396/

U.S. Army Audit Agency, "Army Rail Operations—Captive Fleet," Audit Report A-2017-0023-ALS, January 10, 2017.

U.S. Army Combat Readiness/Safety Center, "Command Outbrief, Railcar Derailment," PowerPoint briefing, July 13, 2010.

U.S. Army Fort Bragg, "82nd Airborne Division," undated. As of December 1, 2016:
https://www.bragg.army.mil/index.php/units-tenants/82nd-airborne-division

U.S. Military Surface Deployment and Distribution Command, "Anchor-Hole Flatcar Chain Distribution," memorandum, undated, circa 2014.

U.S. Military Surface Deployment and Distribution Command, Transportation Engineering Agency, "Tiedown Instructions for Rail Movements," SDDC TEA Pamphlet No. MI 55-19, seventh edition, July 2015. As of March 22, 2017:
https://www.sddc.army.mil/sites/TEA/Functions/Deployability/TransportabilityEngineering/Transportability%20Engineering%20Publications/MI_55-19_7th_edition.pdf

Weisgerber, Marcus, "Rail Issue: U.S. Military Needs New Way to Ship Heavy Vehicles," *Defense News*, April 1, 2013, p. 18. As of November 29, 2016:
http://search.proquest.com/docview/1326743403?accountid=25333